BACTERIAL TRANSFORMATION

European Meeting on Bacterial Transformation, 1st, Gulbenkian Institute of Science, 1972.

BACTERIAL TRANSFORMATION

Proceedings of the First European Meeting
held at the Gulbenkian Institute of Science, Oeiras, Portugal
on August 31 — September 2, 1972

Edited by

LUÍS J. ARCHER

Laboratory of Microbiology
Gulbenkian Institute of Science
Oeiras, Portugal

1973

ACADEMIC PRESS · LONDON AND NEW YORK

ACADEMIC PRESS INC. (LONDON) LTD.
24/28 Oval Road,
London NW1 7DX

United States Edition published by
ACADEMIC PRESS INC.
111 Fifth Avenue
New York, New York 10003

Library of Congress Catalog Card Number: 73-2288
ISBN: 0 12 059450 1

PRINTED IN GREAT BRITAIN BY
J. W. ARROWSMITH LTD., BRISTOL

PREFACE

The First European Meeting on Bacterial Transformation was originally planned as a small and informal gathering of scientists actively engaged in Transformation research.

The location and resources of the Gulbenkian Institute of Science made possible a Meeting, which not only was pleasant but had a rather wider representation than had originally been envisaged.

This fact, together with the quality of the Communications, persuaded us to publish the present volume. A word of appreciation should be given to Dr. Frank E. Young for his initial suggestion and enthusiastic support for this decision.

We only regret that a number of brilliant communications could not be included in this volume. These were papers given by Drs. C.Anagnostopoulos; Hideo Hirokawa; S.G. Hughes and John F. Collins; Corinne Jamet - Vierny and C. Anagnostopoulos; Carmen G. Leonard and Roger M. Cole ; Max Mergeay; Roman Pakula; Thomas Trautner.They have already been or will soon be published elsewhere.

Special gratitude should be expressed to Dr. Otto E. Landman and Dr. Constantine Anagnostopoulos for the help and encouragement they gave from the earliest stages of the organization of the Meeting to its successful conclusion.

A word of appreciation should also be given to Mrs. La-Salette Costa for the expertise and the spirit of dedication and sacrifice with which she typed this book.

Laboratório de Microbiologia
Instituto Gulbenkian de Ciência
Oeiras, Portugal.

Luís J. Archer

CONTENTS

next

III - RECOMBINATION

Bacillus subtilis

Pneumococcus and Streptococcus

EARLY STEPS IN TRANSFORMATION

ANALYSIS OF STEPS IN TRANSFORMATION AND TRANSFECTION IN TRANSFORMATION-DEFECTIVE MUTANTS AND LYSOGENIC STRAINS OF BACILLUS SUBTILIS

R. E. YASBIN
M. J. TEVETHIA†
G. A. WILSON
F. E. YOUNG

Department of Microbiology
The University of Rochester
Rochester, New York

ABSTRACT

Two approaches were utilized to dissect the steps in DNA mediated transformation in Bacillus subtilis. First mutants were isolated with blocks at different stages in transformation. The most extensively studied strain, RUB19 was unable to be transformed by denatured and double stranded bacterial DNA or transfected by DNA isolated from bacteriophage SP01. It was transfected at reduced levels by DNA isolated from bacteriophage Ø29. RUB19 has increased sensitivity to U.V. irradiation, mitomycin C and methylmethane sulfonate. Although increased sensitivity is suggestive of a deficiency in the ability of this mutant to undergo genetic recombination or repair, RUB19 could be transduced by bacteriophages PBS1 and SP10 and could yield recombinant phage genomes when multiply infected with mutant strains of bacteriophage Ø29. In addition, RUB19 carrying prophage Ø105 was unable to release this bacteriophage at wild type levels by either sponta-

† The University of Texas at Houston
M.D. Anderson Hospital and Tumor Institute and
Graduate School of Biomedical Sciences at Houston, Texas

neous or mitomycin C induction while RUB19 carrying prophage SP02 was able to release this bacteriophage following induction at wild type frequency.

Because of the observed decrease in excision of Ø105 in RUB19, and the previous observation that the establishment of the lysogenic state in *B.subtilis* also depresses the level of transformation, we also examined the effect of lysogeny on transformation and transfection. Although strains carrying SP02 or Ø105 have reduced levels of transformation,and superinfection marker rescue,lysogenic strains were unaltered in : (1) the frequency of transfection with DNA isolated from bacteriophages Ø29 and SP01; (2) the ability to be transduced by bacteriophages PBS1 and SP10; (3) the sensitivity to U.V. irradiation ; and (4) the ability to produce genetically recombined phage progeny.Therefore,some steps in transformation and transfection can be clearly separated in lysogenic strains and poorly transformable mutants.

1. INTRODUCTION

The metabolic processes regulating the recombination of genetic information are complex.For instance, Clark postulated the existence of at least twelve different genes involved in recombination in *Escherichia coli* (Clark, 1971). In order to study these reactions,mutants have been isolated that are deficient in their ability to exchange deoxyribonucleic acid (DNA). Selection in *E. coli*, the most extensively investigated model system, was based on : (1) sensitivity to x-rays (Emmerson, 1968; Willetts & Mount,1969);(2) reduced production of bacteriophage in lysogenic cultures (Clark, 1967;Hertman & Luria, 1967); or (3) decrease in genetic exchange (Clark, 1971).In order to define the recombination processes in *Bacillus subtilis*, we first isolated putative *Rec*$^-$ mutants.Because previous studies have shown that lysogens of *B.subtilis*,*E.coli* and streptococci are inhibited in their ability to successfully complete genetic exchange (Arber, 1967; Peterson & Rutberg, 1969; Yasbin & Young, 1972; Parsens, Colon, Ranhand, Leonard & Cole, Abst.Ann.Meet.Amer. Soc. Microbiol. p. 50, 1972), we chose to analyse the effects of lysogeny on transformation. Therefore mutants and lysogenic strains that were unable to be transformed were constructed and characterized according to their ability to recombine genetic material. In addition to transformation, the *B. subtilis* model system has three other procedures which can be used to demonstrate genetic recombination;transduction (Hoch, Barat & Anagnostopoulos, 1967 ; Dubnau and Davidoff-Abelson, 1969 ; Hoch & Anagnostopoulos, 1970); transfection with mature bacteriophage DNA (Okubo, Strauss & Stodolsky, 1964); and transfection with prophage DNA (Rutberg, Hoch & Spizizen, 1969; Rutberg & Rutberg,1971).In this study the influence of mutations and lysogeny were assayed with these different methods.

2. MATERIALS AND METHODS

(a) Bacterial strains and bacteriophages

Strain BR151(*lys*-3,*trp*C2, *met*B10)and strain *B.subtilis* H were obtained from B. Reilly. Strains RUB19, RUB39, and RUB67 are nontransformable mutants of BR151 isolated following treatment with N-methyl-N'nitro-nitrosoguanidine (NTG). *B. subtilis* 168 and bacteriophages SP02 and SP02c1 were obtained from W.Romig and bacteriophages Ø105, Ø29 and SP01 were received from J. Hoch, B. Reilly and K. Bott, respectively. The *rec*B mutant was strain MB209 of the Marmur collection. The temperature-sensitive (*ts*) mutants of Ø29 were obtained from D.Anderson. All other bacteria and bacteriophages utilized in addition to the procedures for the growth and propagation of the bacteria and bacteriophages have been previously described (Yasbin & Young, 1972 ; Yasbin, Wilson & Young, J. Bacteriol., in press).

(b) Procedures for genetic exchange

Except for specified experiments, the procedures for isolation of DNA, transformation, transfection, and binding of radioactive DNA to competent cells are those described previously (Boylan, Mendelson, Brooks & Young, 1972 ; Yasbin & Young, 1972; Yasbin, Wilson & Young, J. Bacteriol., in press).Competent cells were transformed with denatured DNA according to the procedure of Tevethia *et al*. (1970, 1971).

(c) Isolation of nontransformable mutants

Exponentially growing cultures of BR151 were treated with NTG using the procedure of Adelberg *et al*. (1965). Isolated colonies were tested for their ability to be transformed and transfected.

(d) Transduction and recombination procedures

Transduction with PBS1 and SP10 were performed by standard procedures (Lovett & Young, 1970) and (Young, Smith & Reilly, 1969) on the same selective media that were utilized for transformation.Recombination of the *ts* mutants of Ø29 involved a modification of the procedure of Hagen *et al*. (1971).An exponentially growing culture (3 ml with a turbidity of 50 Klett units) in Penassay broth(Difco Antibiotic Medium III) was incubated for 30 min at 4° C with each of the <u>ts</u> mutants at a multiplicity of infection (moi) of 5.The infected cells were then incubated with aeration at 37° C for 10 min, filtered (Millipore HAWP 025, HA 0.45μ), washed, resuspended in Penassay broth, and incubated at 37° C until lysis occurred. The lysates were diluted and the bacteriophages were titered. Total bacteriophage were determined at 37°C and the number of wild-type recombinants were those that formed plaques at 45°C.

(e) Ultraviolet irradiation of cells

Strain BR151 and nontransforming mutants were grown on Tryptose Blood Agar Base (Difco) for 24 hr at 37°C before being suspended in 6 ml of Spizizen's minimal salts (Anagnostopoulos & Spizizen, 1961) to a final concentration of 10^8 cells per ml for irradiation. The cells were irradiated (8 watt General Electric Germicidal lamp) at a rate of 10 erg/mm^2 per sec., diluted, and plated on TBAB. Competent populations of strains 168, 168(Ø105) and MB209 were suspended in Spizizen minimal salts, irradiated, diluted and plated on Spizizen's minimal agar+0.5% DL-tryptophan in an overlay of soft agar containing 2% casein hydrolysate.

(f) Induction of prophages

Bacteria were grown in Penassay to turbidity of 50 Klett units, incubated for 15 min with 0.5μg/ml of mitomycin C (Calbiochem), pelleted by centrifugation(1,200 x g), washed, and resuspended in Penassay broth. Samples were taken immediately following resuspension and after 3 hr of incubation at 37° C with aeration. Phage were assayed on strain BR151 with an overlay of M agar without glucose (Yehle & Doi, 1967).

3. RESULTS

(a) Isolation and characterization of nontransformable mutants

Unlike the Haemophilus system (Caster, Postel & Goodgal, 1970; Beattie & Setlow, 1971), there is presently no available procedure for the primary selection of nontransformable (Tfm^-) strains in *B. subtilis*. Each isolated colony must be tested initially for its ability to undergo DNA-mediated transformation. Once isolated, the nontransformable mutants are phenotypically characterized as to their ability to bind radioactive DNA, to be transformed with native and denatured DNA, and to be transfected with DNA isolated from bacteriophages Ø29 and SP01. In Table I the results of these tests on three nontransformable mutants and their competent parent strain (BR151) are shown. None of the mutants could be transformed with either native or denatured DNA nor could they bind more than 2% of the radioactive DNA bound by strain BR151. The mutant strain RUB19 was able to be transfected at reduced levels with DNA from bacteriophage Ø29 but none of the mutants could be successfully transfected with DNA from bacteriophage SP01. Because strains RUB39 and RUB67 were unable to bind DNA, they were most likely deficient in the absorption and uptake phases of transformation (Tomasz, 1969). While capable of transfection, strain RUB19 also did not bind any appreciable amount of radioactive DNA. A possible explanation of these conflicting results might be an alteration in the competence cycle for strain RUB19. Such an alteration would establish a new time

TABLE I

TRANSFORMATION AND TRANSFECTION OF BR151 AND TFM$^{-\dagger}$ MUTANT STRAINS

Recipient Strain	Trp^+ Transformants per 10^8 cells: Native DNA	Trp^+ Transformants per 10^8 cells: Denatured DNA	Irreversible Binding of DNA cpm‡	Transfection pfu per 10^8 cells: Ø29 DNA	Transfection pfu per 10^8 cells: SP01 DNA
BR151	1.0×10^6	6.1×10^4	4762.0	1.6×10^5	2.8×10^5
RUB19	$<10^1$	$<10^1$	69.1	2.5×10^3	$<10^1$
RUB39	$<10^1$	$<10^1$	69.0	$<10^1$	$<10^1$
RUB67	$<10^1$	$<10^1$	82.5	$<10^1$	$<10^1$

† nontransformable mutants

‡ counts per minute

for the development of maximal competence. To investigate this possibility, transfection of strain RUB19 was studied over a 3 hr period. Because both strains BR151 and RUB19 achieved their maximum frequencies of transfection at the same time (Fig. 1), it is unlikely that they differ significantly in the temporal appearance of maximal competence. Note that there was a marked difference in the maintenance of transfection.

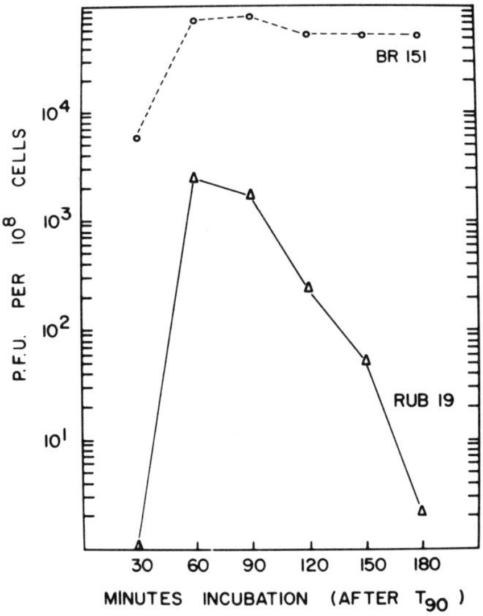

Fig. 1 - The transfection of strains BR151 and RUB19 during the three hour period following the dilution of the competent cells into SP11 medium (T_{90}, ninety minutes after the cessation of exponential growth in SP1 medium). DNA isolated from bacteriophage Ø29 was incubated with the bacteria as described in Materials and Methods. Plaque forming units determined after transfection on indicator strain *B. subtilis* H; O--O , BR151; △—△, RUB19.

(b) Recombination in RUB19

Transfection with DNA from bacteriophage SP01 has been shown to require genetic recombination or repair (Okubo, Strauss & Stodolsky, 1964), while DNA from bacteriophage Ø29 has no such requirement (Reilly & Spizizen,1965).The ability of strain RUB19 to be transfected with DNA from Ø29 but not from SP01 suggested that this strain lacked a recombination or repair system. This contention was further supported by the increased sensitivity of mutant RUB19 to ultraviolet irradiation as compared with strains BR151, RUB39 and RUB67 (Fig. 2). Strain RUB19 was also more sen-

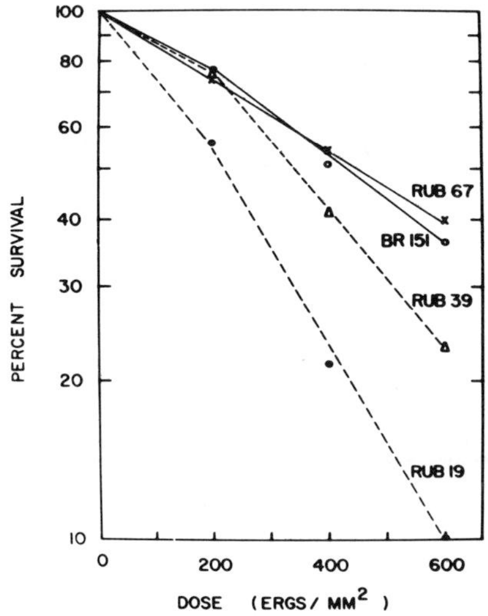

Fig. 2 - The sensitivity to UV irradiation of strains: BR151, O--O ; RUB19, ●--●; RUB39, △--△; RUB67, x-x.

sitive to the killing effects of mitomycin C and methyl-methane sulfonate (unpublished observations). Although these increased sensitivities are usually indicative of nonfunctional recombination or repair systems, strain RUB19 was transduced by both PBS1 and SP10 and was capable of recombining two _ts_ mutants of Ø29 at the same frequencies obtained with strains BR151,RUB39 and RUB67 (Table II) .These results, however,do not eliminate the possibility that strain RUB19 is deficient in a recombination system that is not essential for transduction or for recombination of bacteriophages.

TABLE II

RECOMBINATION IN BR151 AND TFM$^{-\dagger}$MUTANT STRAINS

Bacterial strains	Trp^+ Transduction per ml PBS1	SP10	% Recombination of Ø29 mutants
BR151	4,700	1,630	6.0
RUB19	3,000	2,730	9.0
RUB39	3,800	1,950	5.0
RUB67	4,100	1,610	15.0

†nontransformable mutants

(c) Inability of a _Rec_$^-$ mutant to release prophages

The release of prophage from lysogenic bacteria has been used as an index of the _Rec_$^-$ phenotype. For instance, _recA_ mutants are deficient in induction of bacteriophages in _E. coli_ and _B. subtilis_ (Clark, 1967; Hertman & Luria, 1967; Rutberg & Rutberg, 1971).To investigate the effect of the Tfm$^-$ mutation in RUB19 on induction, the strain was incubated with bacteriophages

Ø105 and SP02 and lysogenic clones were isolated. Although bacteriophage Ø105 was not induced at normal frequencies, strain RUB19 was able to release the temperate bacteriophage SP02 at frequencies equivalent to those from BR151(Table III).This difference can be explained either by the different dependencies that the two temperate bacteriophages have on the host enzyme systems (Rutberg, Armentrout & Jonasson, 1972)or by an alteration of the integration site for bacteriophage Ø105.

TABLE III

INDUCTION OF Ø105 AND SP02

Bacterial strain	Viable count per ml	Induced cells per ml	Induction %
BR151 (Ø105)	3.22×10^8 2.76×10^8	3.82×10^8 3.17×10^8	116
BR151 (SP02)	2.78×10^8 2.65×10^8	3.4×10^8 1.98×10^8	99
RUB19 (Ø105)	1.80×10^8 2.64×10^8	3.0×10^5 1.83×10^6	0.47
RUB19 (SP02)	2.40×10^8 2.68×10^8	2.55×10^8 2.90×10^8	107

(d) Effect of lysogeny on transformation and transfection in *B. subtilis*

The establishment of the lysogenic state in competent cultures of *B.subtilis* reduced the ability of the bacteria to be transformed and to bind bacterial DNA (Peterson & Rutberg, 1969; Yasbin & Young, 1972 ; Yasbin, Wilson & Young, J. Bacteriol., in press). However, these lysogenic cultures were not inhibited in their ability to be transfected by DNA isolated from bacteriophages Ø29 and SP01. This reduction in transformation occurred both with native and denatured DNA(Table IV). Because modification

TABLE IV

TRANSFORMATION WITH NATIVE AND DENATURED DNA

Recipient strain	DNA	EDTA†	Trp^+ transformants per ml‡
168	Native§	-	1.54×10^5
	Denatured*	+	8.80×10^4
168(Ø105)-1	Native	-	4.0×10^1
	Denatured	+	1.5×10^1
168(Ø105)-2	Native	-	1.55×10^2
	Denatured	+	1.00×10^2

†Ethylenediamine tetraacetic acid at final concentration of 1×10^{-3} M when present.

‡Transformants per milliliter per 0.1μg DNA isolated from the prototrophic W23 strain.

§Double stranded DNA.

*Single stranded DNA.

of the chromosome could occur following lysogeny, lysogenic recipients were incubated with DNA isolated from various strains carrying prophages Ø105, SPO2, or both Ø105 and SPO2. As shown in Table V, the lysogeny of donor strains did not overcome the reduction of transformation in lysogenic recipient cells.

Nonlysogenic cells maintained maximal frequencies of transfection and transformation for a period of three hours (Fig. 3). In the lysogenic cultures, however, the frequencies of

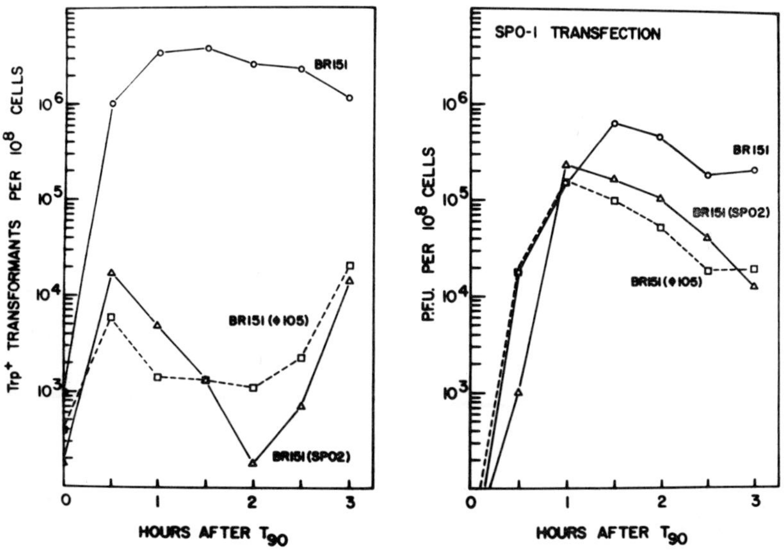

Fig. 3 - The transformation and transfection of strains BR151, BR151 (Ø105) and BR151 (SPO2) during the three hour period following dilution of the competent cells into SP11 medium (T_{90}, ninety minutes after the cells had ceased exponential growth in SP1medium). DNA isolated from bacterial strain RUB818 or bacteriophage SPO1 was incubated with competent cells as described in Materials and Methods. O—O, BR151; □--□, BR151(Ø105), △—△, BR151 (SPO2).

TABLE V

EFFECT OF LYSOGENY ON THE EFFICIENCY OF DONOR DNA

Experiment	Donor DNA	Recipient Strain	Trp^+ Transformants per 10^8 cells	Relative† Transformation
1	RUB818	BR151	3,500,000	1.0
	0.3 μg/ml‡	BR151(SPO2)	67,000	0.019
		BR151(Ø105,SPO2)	360	0.0001
2	RUB818(SPO2)	BR151	3,000,000	1.0
	0.13 μg/ml	BR151(SPO2)	43,000	0.014
		BR151(Ø105,SPO2)	470	0.0001
3	RUB818(Ø105,SPO2)	BR151	1,800,000	1.0
	0.08 μg/ml	BR151(SPO2)	30,000	0.017
		BR151(Ø105,SPO2)	390	0.0002

†Relative transformation determined by dividing the average frequency of transformation for each sample by the average frequency of transformation for BR151.

‡Final concentration of DNA in transformation mixture.

transfection (for SP01)are comparable to the wild type frequency at 60 min but decline more rapidly while the frequencies of transformation never achieve the levels obtained in the non-lysogenic strains.Unlike the parental strain, the lysogenic cultures have two periods of maximal transformation, one at 30 min and another after 240 min. Thus, the time of addition of the DNA is extremely important when comparing the frequencies of transformation and transfection in the lysogenic cultures.

(e) Superinfection marker rescue and recombination

The abrupt shutoff of transfection of lysogenic cultures with DNA from a bacteriophage which requires a recombination or repair system (Spatz & Trautner, 1971) suggested that lysogeny may alter one of the enzymes responsible for recombination of fragments of DNA. To investigate this possibility, lysogenic and nonlysogenic cultures were tested for their sensitivity to UV irradiation and for their ability to mediate superinfection marker rescue.

As previously discussed, many of the $\underline{Rec}^-$ mutants isolated are sensitive to ultraviolet irradiation.Competent cultures of 168,168(Ø105)and MB209 were irradiated and their sensitivities were determined (Fig. 4). While recB shows an extreme sensitivity to UV irradiation, no difference appears between the sensitivities of the lysogenic and nonlysogenic cultures.

To determine if cultures were capable of mediating superinfection marker rescue,lysogenic and nonlysogenic recipients were incubated with double and single-stranded DNA from a clear plaque mutant of bacteriophage SP02(SP02cl). After 20 min of in-

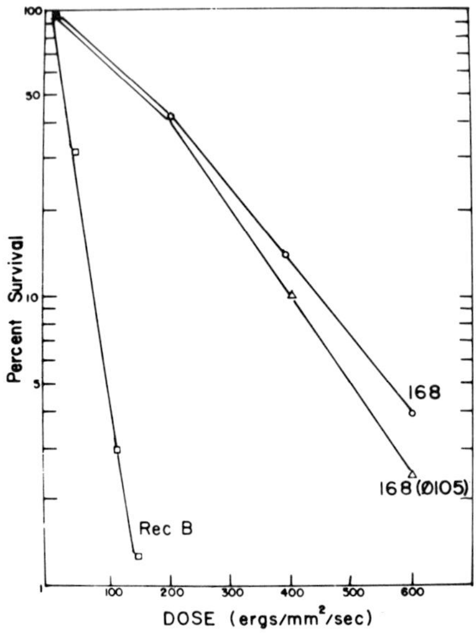

Fig. 4 - The sensitivity to UV irradiation of competent strains of B. subtilis 168, O–O; 168(Ø105), Δ–Δ; and MB209 carrying recB, □–□

cubation at 37° C,the cultures were diluted into either Penassay broth (Table VI;control) or mixed with Penassay broth containing a suppressible mutant SP02B at a moi of 20(Table VI;experimental). Infectious centers were assayed on a strain of B. subtilis 168 that was resistant to bacteriophage Ø105 and that could not be infected by SP02B. As shown in Table VI,transfection with native DNA from SP02cl is more efficient in lysogenic recipients. Upon superinfection, the transfection increases from 6.2×10^3 pfu/ml to 1.8×10^5 pfu/ml in the nonlysogen, and remains essentially constant in the lysogen.A more profound effect of superinfection marker rescue (SMR) is demonstrated in transfection experiments with denatured DNA.As noted earlier by Dishman (1972),B.subtilis cannot be transfected by denatured DNA from bacteriophage SP02.

TABLE VI

SUPERINFECTION MARKER RESCUE OF SPO2†

Recipient strain	DNA	Transfectants per ml Control	Experimental
168	Native	6.2×10^3	1.8×10^5
168(Ø105)	Native	1.8×10^5	1.3×10^5
168	Denatured	$<4 \times 10^1$	1.1×10^5
168(Ø105)	Denatured	$<4 \times 10^1$	2.3×10^3

†Competent cells were incubated with DNA isolated from SPO2cl. After 20 min the reaction was terminated with DNase, and divided into two aliquots. Penassay broth (0.1 ml) was added to the control while the experimental tube received 0.1 ml of SPO2B(moi=20). Transfectants were scored on a lawn of *B. subtilis* 168 carrying *trp*C2 and resistant to Ø105.

However, this DNA can enter the cell and give rise to transfectants following SMR. Note that SMR is 50-fold lower in the lysogenic strain than in the nonlysogens. This observation suggests that lysogens may be deficient in one of the recombination processes.

4. DISCUSSION

(a) Analysis of transformation and transfection

The process of transformation in *B. subtilis* has been divided into the following stages: (1) the attachment of exogenous

DNA to a competent cell (Young & Spizizen, 1961 ; Tomasz, 1969; Erickson, 1970); (2) the formation of a DNA-cell complex that is resistant to deoxyribonuclease (DNase) (Young, 1967; Dubnau & Cirigliano, 1972 a, b) ; (3) the formation of a DNA-cell complex where the DNA has single-stranded characteristics(Erickson,Young & Braun, 1969 ; Erickson, 1970; Dubnau & Cirigliano, 1972b); (4) the transport of the donor DNA across the cell membrane; (5) the establishment of a donor recipient complex (Dubnau & Davidoff-Abelson, 1971); (6) the integration of the donor DNA into the recipient chromosome; and,(7) the replication of the recombinant. Characterization of nontransformable mutants based on the nature of their defects could be accomplished by testing each mutant for its ability to bind radioactive DNA and to be transformed by denatured (single-stranded) DNA. Strains that are unable to bind DNA or to render the DNA insensitive to DNase demonstrate little or no binding of radioactive DNA. In *Haemophilus*,nontransformable mutants have been recognized that could be transformed with denatured but not native DNA(Caster, Postel & Goodgal, 1970).These mutants are deficient in the metabolic processes which convert native donor DNA to its single stranded form. Transfection in *B. subtilis* should also provide an index of a cell's ability to utilize exogenous DNA.Theoretically, transfection should include the stages of transformation except those associated with the pairing and integration of the donor DNA (recombination processes). Thus,a nontransformable mutant capable of utilizing transfecting DNA should, *a priori*, be deficient in a recombination process.This is undoubtedly an oversimplification of the differences between transformation and transfection as evidenced by the following observations.Transfection has been shown in certain cases to require genetic recombination or repair (Okubo, Strauss & Stodolsky,1964;Rutberg, Hoch & Spizizen, 1969;Rutberg & Rutberg,

1971; Spatz & Trautner, 1971). As previously discussed, the mutation in strain RUB19 and lysogeny affect transformation and transfection differently. In addition, Oostindier-Braaksma and Epstein (1970) have shown that the period of maximal competence for DNA from bacteriophage SP82 is later than that for bacteriophage SP02 and bacterial DNA. High concentrations of nalidixic acid are inhibitory for transformation and transfection with DNA from bacteriophage SP01 but not with DNA from bacteriophage Ø29 (Wilson and Bott, 1971). These data indicate that differences exist not only between the processes of transformation and transfection but also in the different transfecting systems. Therefore, nontransformable mutants capable of DNA mediated transfection are not necessarily recombination deficient. Nevertheless, transfection remains a good method for screening nontransformable mutants for potential recombination deficient strains.

(b) Binding of radioactive DNA to competent cells

In general, the binding of radioactive DNA can serve as an indication of the level of competence (Young & Spizizen, 1961). However, the binding of DNA and transformation can be disassociated under adverse conditions (Young & Spizizen, 1963). All three of the nontransformable mutants (RUB19, RUB39, RUB67) bind less than 2% of the radioactive DNA bound by competent strain BR151. These results would suggest that a correlation exists between binding of DNA and the level of competence if it were not for the paradoxical results observed with strain RUB19. Although strain RUB19 bound no more radioactive DNA than did strains RUB39 or RUB67, it was able to be transfected by DNA from bacteriophage Ø29. In addition, lysogenic cultures of _B. subtilis_ lose their ability to be transformed for genetic markers prior to their loss of ability to bind radioactive DNA (Yasbin, Wilson & Young, J. Bacteriol., in press). These same lysogenic cells are capable of normal levels of transfection while they are only able to bind extremely small amounts of radioactive DNA. Thus in at least three instances, changes in the ability of cells to be transformed or transfected did not result in parallel changes in the ability of these cells to bind radioactive DNA. These results can be explained if one of the following hypotheses are correct: (1) Most of the radioactive DNA bound by competent cells is superfluous and not used in transformation and transfection. (2) Donor DNA that is bound to competent cells but not transported across the membrane is eventually degraded. Therefore, cells that are unable to integrate or to transport donor DNA would appear to be deficient in binding of DNA. (3) The nontransformable mutants and lysogens differ from the parental strain in the rate of inactivation of donor DNA isolated from bacteria and bacteriophages. (4) Bacterial DNA and bacteriophage DNA are transported by different mechanisms in competent cells. Presently it is not possible to decide between these alternatives.

(c) Recombination of deficient strains

Only half of the 12 genes postulated for processes associated with recombination (Clark, 1971) have been identified in *E. coli*. In *B. subtilis* three *Rec*$^-$ strains have been isolated and partially characterized (Hoch, Barat & Anagnostopoulos, 1969; Dubnau, Davidoff-Abelson & Smith, 1969; Rutberg & Rutberg, 1971). These studies have shown that transformation frequencies are drastically reduced in *rec*A and *rec*4 strains and less severely reduced in recB. Transduction with bacteriophage SP10 is reduced in Rec$^-$ strains while transduction with bacteriophage PBS1 occurs at essentially wild type frequencies. The recA gene is needed for transfection mediated by prophage DNA while the recB gene is necessary for marker rescue from mature bacteriophage DNA. Recombination of genetic material has not been demonstrated in rec4 strains. These data indicate that the number of genes regulating recombination in B. subtilis are probably as numerous as in E. coli. Nontransformable mutants and lysogenic cultures represent a possible mechanism for the elucidation of the various recombination systems in B. subtilis. Evidence for recombination in strain RUB19 was provided by its ability to be transduced by bacteriophages PBS1 and SP10, to recombine bacteriophage DNA, and to be transfected at reduced levels by bacteriophage Ø29 DNA.

However, its inability to be transfected by DNA from bacteriophage SP01 or to be transformed by bacterial DNA as well as its sensitivity to UV irradiation suggests that this is a recombination deficient strain. Similarly, the inability of lysogens to mediate SMR is also indicative of a recombination deficiency. These deficiencies may involve recombination systems which only regulate specific genetic exchanges. Alternatively, during the physiological stage of competence, recombination systems may be utilized which are not essential at other stages of growth. Bacteria could then be isolated which would be unable to recombine genetic material during only one physiological stage.

Although a clear characterization of the genetic defects in strains RUB19, RUB39, RUB67 and lysogens of B. subtilis have not been completed, two conclusions have resulted : (1) greater differences exist between the processes of transformation and transfection than merely the integration of genetic material into a recipient chromosome, and (2) the significance of radioactive binding of DNA to competent cells must be re-evaluated.

ACKNOWLEDGEMENTS

This study was aided by grants from the American Cancer Society (VC-271), the National Institute of General Medical Sciences (5-T01-GM00592-12),and for one of us (MJT) Robert A. Welch Foundation(G474).A preliminary report of this work was presented at the 1972 annual meeting of the American Society for Microbiology.

5. REFERENCES

ADELBERG, E. A., MANDEL, M. & CHEN, G. C. C. (1965). Biochem.Biophys. Res. Commun. 18:788.

ANAGNOSTOPOULOS, C. & SPIZIZEN, J. (1961). J. Bacteriol.81:741.

ARBER, W. (1965). Ann. Rev. Microbiol. 19:365.

BEATTIE, K. L. & SETLOW, J. K. (1971). Nature N. B. 231:177.

BOYLAN, R. J., MENDELSON, N. H., BROOKS, D. & YOUNG, F. E. (1972). J. Bacteriol. 110:281.

CASTER, J. H., POSTEL, E. H. & GOODGAL, S. H. (1970). Nature 227:515.

CLARK, A. J. (1967). J. Cell Physiol. 70, suppl. 1:165.

CLARK, A. J. (1971). Ann. Rev. Microbiol. 25:437.

DISHMAN, B. R. (1972). Thesis, Univ. Texas, Graduate School Biomedical Sciences at Houston.

DUBNAU, D. & CIRIGLIANO, C. (1972a). J. Mol. Biol. 64:9.

DUBNAU, D. & CIRIGLIANO, C. (1972b). J. Mol. Biol. 64:31.

DUBNAU, D. & DAVIDOFF-ABELSON, R. (1971). J. Mol. Biol. 56:209.

DUBNAU, D., DAVIDOFF-ABELSON, R. & SMITH, I. (1969). J. Mol. Biol. 45:155.

EMMERSON, P. T. (1968). Genetics 60:19.

ERICKSON, R. J. (1970). Curr. Top. Microbiol. and Immunol. 53:149.

ERICKSON, R. J., YOUNG, F. E. & BRAUN, W. (1969). J. Bacteriol. 99:125.

HAGEN, E. W., ZECCE, V. M. & ANDERSON, D. L. (1971). Virology 43:561.

HERTMAN, I. & LURIA, S. E. (1967). J. Mol. Biol. 23:117.

HOCH, J. A. & ANAGNOSTOPOULOS, C. (1970). J. Bacteriol. 103:295.

HOCH, J. A., BARAT, M. & ANAGNOSTOPOULOS, C. (1967). J. Bacteriol. 93:1925.

LOVETT, P. S. & YOUNG, F. E. (1970). J. Bacteriol. 101:603.

OKUBO, S., STRAUSS, B. & STODOLSKY, M. (1964). Virology 24:552.

OOSTINDIER-BRAAKSMA, F. & EPSTEIN, H. T. (1970). Molec. Gen. Genet. 108:23.

PETERSON, A. & RUTBERG, L. (1969). J. Bacteriol. 98:874.

REILLY, B. E. & SPIZIZEN, J. (1965). J. Bacteriol. 89:782.

RUTBERG, L., ARMENTROUT, R. W. & JONASSON, J. (1972). J. Virol. 9:732.

RUTBERG, L., HOCH, J. A. & SPIZIZEN, J. (1969). J. Virol. 4:50.

RUTBERG, B. & RUTBERG, L. (1971). J. Virol. 8:919.

SPATZ, H. C. & TRAUTNER, T. A. (1971). Mol. Gen. Genet. 113:174.

TEVETHIA, M. J. & CAUDILL, C. P. (1971) J. Bacteriol. 106:808.

TEVETHIA, M. J. & MANDEL, M. (1970). J. Bacteriol. 101:844.

TOMASZ, A. (1969). Ann. Rev. Genetics 3:217.

WILLETTS, N. S. & MOUNT, D. W. (1969). J. Bacteriol. 100:923.

WILSON, G. A. & BOTT, K. F. (1971). Informative Molecules in Biological Systems, L. G. H. Ledoux ed., North Holland pub., pg. 38.

YASBIN, R. E. & YOUNG, F. E. (1972). Biochem. Biophys. Res.Commun. 47:365.

YEHLE, C. O. & DOI, R. H. (1967). J. Virol. 1:935.

YOUNG, F. E. (1967). Nature 213:773.

YOUNG, F. E., SMITH, C. & REILLY, B. E. (1969). J. Bacteriol. 98:1087.

YOUNG, F. E. & SPIZIZEN, J. (1961). J. Bacteriol. 81:823.

YOUNG, F. E. & SPIZIZEN, J. (1963). J. Bacteriol. 86:392.

GENETICAL AND BIOCHEMICAL CHARACTERIZATION OF *BACILLUS SUBTILIS* MUTANTS ALTERED IN TRANSFORMATION

MARIO POLSINELLI
GIORGIO MAZZA
UMBERTO CANOSI
ARTURO FALASCHI

Istituto di Genetica, Università di Pavia,
and
Laboratorio di Genetica Biochimica ed Evoluzionistica
del Consiglio Nazionale delle Ricerche
27100 Pavia, Italy

ABSTRACT

We have isolated mutants of *Bacillus subtilis* defective in transformation, with the following selection criteria: inability to be transformed on solid medium at 37°C; inability to be transformed on solid medium at 42°C; sensitivity to ultraviolet light; sensitivity to methyl-methane-sulfonate. The isolated mutants have been characterized for a number of properties, including mitomycin sensitivity, ability to be transformed and transfected, DNA polymerase activity. The most interesting mutants were the following: one reduced in transformation more than in transfection; one markedly thermosensitive in the process of transformation but with normal ability to take up DNA; one which is deficient in DNA polymerase I.

1. INTRODUCTION

Mutants with impaired transformability and/or ability to repair damage to DNA have been described in *Bacillus subtilis* by several authors (Searashi and Strauss, 1965; Okubo and Romig, 1966; Hoch et al., 1967; Zadrazil and Fucik,1971; Yasbin et al., 1971). Another mutant described by Gass et al.(1971) had some properties consistent with those of the pol AI strain of *Escherichia coli* (De Lucia and Cairns, 1969). In this report we describe the properties of a number of mutants of *B. subtilis* isolated in our laboratory as deficient in transformation and/or repair of damage to DNA.

2. METHODS

The procedure for transformation used was that described by Stewart (1969). Transfection with SPP1 phage DNA was made according to Riva et al.(1968).Transduction with phage PBS1 was performed according to Hoch et al. (1967). Mutagenesis was obtained with nitrosoguanidine(Adelberg et al., 1965). DNA polymerase was assayed according to Okazaki and Kornberg (1964).

3. RESULTS

(a) Mutants defective for transformation

Colonies from mutagenized cultures of strain PB25(his_2, trp_2 , originally SB25 from J. Lederberg) were replica-plated on selective medium spread with trp^+ DNA(50 µg/plate) and incubated at 37°C. Colonies unable to give transformants were considered for further studies. With this procedure we have isolated the five mutants reported in Table I, which were studied also for other properties.

TABLE I

PROPERTIES OF THE MUTANTS SELECTED FOR THE INABILITY TO BE TRANSFORMED AT 37°C

Strain	Transformation a)	Transfection a)	UV sensitivity b)	MMS sensitivity c) (m.i.c. µg/ml)	MIT sensitivity c) (m.i.c. µg/ml)	DNA uptake d)
PB25 (parental)	1	1	1	250	0.05	1
PB1629	0.01	0.05	1	250	0.05	0.40
PB1630	0.10	0.50	1	250	0.05	0.30
PB1631	0.05		1	150	0.05	0.10
PB1625	$<1 \times 10^{-7}$	$<1 \times 10^{-7}$	1	100	0.01	<0.10
PB1632	$<1 \times 10^{-7}$	$<1 \times 10^{-7}$	1	250	0.05	<0.10

a) The data report the frequency, at saturating DNA concentration, of transformation or transfection relative to the parental frequency; this was about $1x10^{-3}$ for transformation and $2x10^{-2}$ for transfection.

b) Measured as frequency of survival on nutrient agar of cells exposed after plating to 375 erg/mm^2 from a germicidal lamp; the survival frequency of the parental was $1x10^{-2}$.

c) Determined as minimum inhibitory concentration after plating on nutrient agar.

d) Determined at 0.05 µg/ml DNA concentration, following DNase treatment and washing by centrifugation.

Transformation was reduced in the mutants to values ranging between 1/100 and less than $1/10^7$ of the wild type value. In all the mutants studied, transfection was also reduced, though not always to the same degree as transformation. Sensitivity to ultraviolet light (UV) was normal in all these mutants. Some of the mutants showed a slight sensitivity to methyl-methane-sulfonate (MMS) and only one to mitomycin (MIT). The DNA uptake was reduced in all of them to different degrees. Two mutants (PB1625 and PB1632), that have completely lost the capacity to be transformed, to be transfected and to take up DNA, are probably altered in the uptake machinery. It is worth noticing that the mutants isolated for transformation deficiency show no or little increase of sensitivity to MMS or UV, whereas, as we shall see later, all the mutants selected for MMS or UV sensitivity are deficient in transformation.

Four other mutants were selected as colonies unable to be transformed on solid medium at 42°C. Their properties are described in Table II. As shown in the table, transformation is reduced also in the parental strain at 42°C by about 50%. All the mutants that were selected at 42°C showed impaired transformability also at 35°C. For three of them (PB1645, PB1648, PB1649) the ratio of transformability at 42°C to that at 35°C is essentially like that of the wild type, so that they can be considered as mutants altered in transformation but not particularly temperature sensitive. On the other hand mutant PB1647 was much more affected at 42°C than at 35°C. Sensitivity to MMS and to MIT is not appreciably altered in all these mutants.

TABLE II

PROPERTIES OF THE MUTANTS SELECTED FOR THE INABILITY TO BE TRANSFORMED AT 42°C[a)]

Strain	Transformation			Sensitivity to MMS (m.i.c. μg/ml)	Sensitivity to MIT m.i.c. μg/ml)
	35°	42°	42°/35°		
PB202 (parental)	1	0.45	0.45	500	0.05
PB1645	0.05	0.01	0.20	500	0.05
PB1648	3.2×10^{-3}	1×10^{-3}	0.31	500	0.05
PB1649	0.05	0.01	0.20	500	0.05
PB1647	0.30	2×10^{-4}	7×10^{-4}	300	0.05

a) The data are given as indicated in the footnotes to Table I.

(b) Further characterization of mutant PB1647, temperature sensitive for transformation

Mutant PB1647 was object of further study to determine the site of the alteration. The mutant grows a little slower at 42°C than at 37°C, but altogether its growth is not severely impaired.

It is known that transformation is a temperature sensitive process also in the wild type strains. We checked therefore the transformation frequency of the parental strain and of the mutant at different temperatures. As shown in Fig. 1, transform-

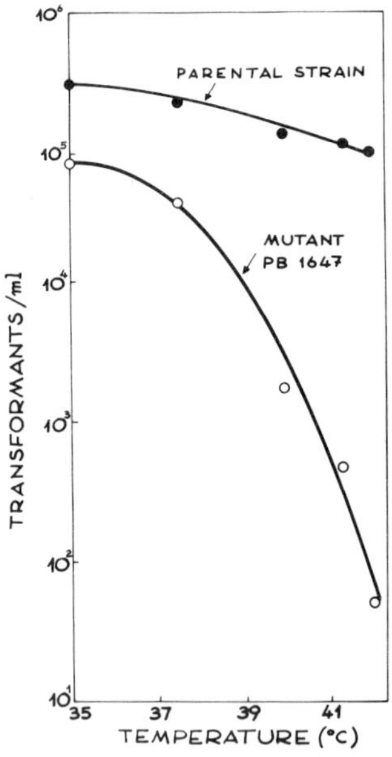

FIG. 1 - TEMPERATURE DEPENDENCE OF TRANSFORMATION FOR STRAIN PB1647

Competent cells of the parental strain PB202 and of strain PB1647 were incubated with prototroph DNA(0.1 μg/ml) for 30 min at the indicated temperatures, and the trp^+ transformants were scored.

ation in the parental strain decreases approximately to one third going from 35°C to 42°C. In the mutant, instead, the decrease of transformation at 42°C is down to one thousandth of the value at 35°C. The viability of both strains was not affected by the same treatment. These data are then good evidence that the mutant is thermosensitive in the process of transformation.

The concentration of DNA strongly affects the transformation frequency of the mutant at high temperatures. This is shown in Fig. 2. A change in DNA concentration between 1 μg/ml and 0.05 μg/ml does not affect the temperature dependence of transformation of the parental strain as shown in the control curve, but

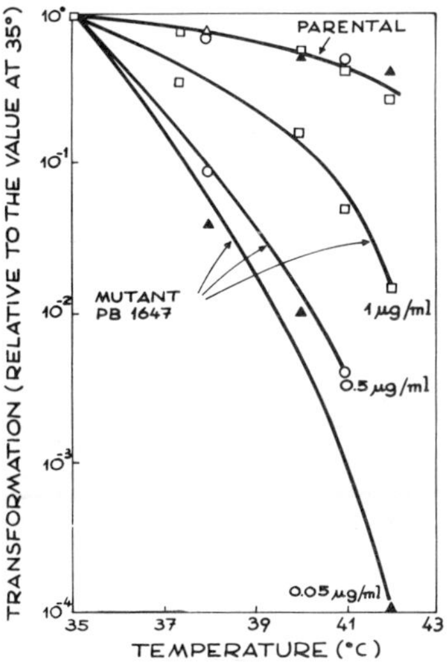

FIG. 2 - TEMPERATURE DEPENDENCE OF TRANSFORMATION OF STRAIN PB1647 AT DIFFERENT DNA CONCENTRATIONS

Experiments as in Fig. 1, but with various DNA concentrations.

enhances the drop in transformation frequency of the mutant at high temperature. This could be an indication that the damaged structure has lost affinity for DNA and needs higher DNA concentrations to give the same effect as in normal conditions. Another possibility is that the alteration in the mutant causes damage to the incoming DNA by activation, for example, of a nuclease; in that case higher DNA concentrations could compensate the damage to DNA.

Does the alteration involve the DNA uptake or a later step? As shown in Fig. 3 the change of temperature between 35°C

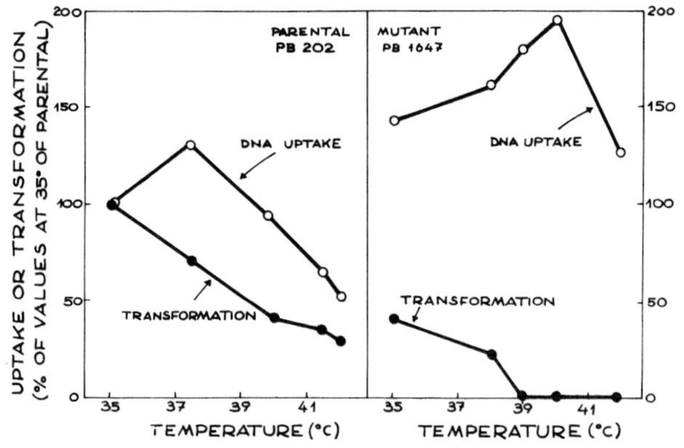

FIG. 3 - TEMPERATURE DEPENDENCE OF TRANSFORMATION OF STRAIN PB1647 AT DIFFERENT DNA CONCENTRATIONS

Experiments as in Fig.1, except that 0.05 µg/ml of ^{3}H labelled DNA was used; the uptake of DNA was measured as reported in Table I.

and 42°C, while reducing transformation in the wild type, has a more complex effect on the uptake also in the parental. In fact, between 35°C and 38°C the uptake rises, while transformation decreases. This rise in uptake takes place also in the mutant, which is not reduced in the ability to take up DNA with respect to wild type at any temperature. It is therefore unlikely that the defect of transformation be due to impaired uptake capacity. Also transfection with SPP1 phage DNA (a process known to require recombination, Spatz and Trautner, 1971; Mazza and Polsinelli, unpublished experiments) is reduced in this mutant at high temperature as shown in Fig. 4. Surprisingly enough, transfection in the parental

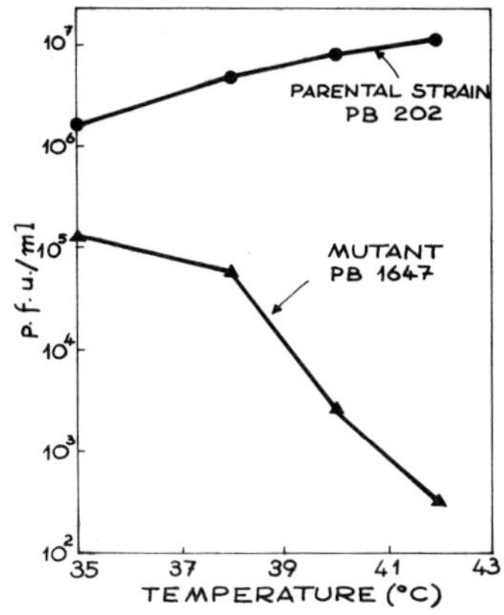

FIG. 4 - TEMPERATURE DEPENDENCE OF TRANSFORMATION OF STRAIN PB1647 AT DIFFERENT DNA CONCENTRATIONS

Competent cells of the parental strain PB202 and of strain PB1647 were transfected with 1 μg/ml SPP1 phage DNA at different temperatures and the infective centers were scored according to the procedure of Riva _et al_. (1968).

strain was increased with rising temperature. This has been observed also by Trautner in the same phage (personal communication). Transfection, therefore, in normal cells seems to respond to temperature quite differently from transformation.

(c) Mutants sensitive to MMS or to UV

A number of mutants sensitive to MMS or UV were isolated from mutagenized cultures of strain PB202(his_2, trp_2, aro_2, tyr_1, originally SB202 from J. Lederberg). The properties of two mutants(PB1641 and PB1642)originally selected for UV sensitivity and of six other isolated as MMS sensitive,are reported in Table III. They have been grouped in three classes according to the criteria reported in the same table. It must be pointed out that all these mutants show impaired transformability except perhaps PB1642 where transformation is reduced only by 30%. DNA uptake however does not seem to be significantly altered in all the mutants.This is as expected if we consider that these strains have been selected for the inability to repair damage to DNA and therefore the alteration should not be involved in the DNA uptake process. It must be also noticed that all the mutants isolated as MMS sensitive are also UV sensitive. Furthermore the degree of sensitivity to UV of the mutants selected as UV sensitive (PB1641 and PB1642) is not appreciably different from that of the mutant selected for MMS sensitivity,as shown in Table III. From one mutant selected for MMS sensitivity (PB1635) and from one selected for UV sensitivity (PB1641),two spontaneous revertants were isolated as MMS resistant(PB1635/1 and PB1641/1).These revertants were then checked for UV sensitivity.As shown in Fig. 5,both revertants had recovered also the original UV resistance. This indicates that the same event impairs at the same time the mechanism of recovery from UV and MMS damages.

(d) Enzymological studies on the UV and MMS sensitive mutants

The mutants isolated for sensitivity to MMS or UV,were also analysed for their DNA polymerase content. Cells from late log cultures of the mutants were washed, sonicated and the crude extracts were assayed for DNA polymerase with native activated calf thymus DNA as template, in the linear range of activity. As shown in Table IV all the mutants have a normal level of activ-

TABLE III

PROPERTIES OF THE MUTANTS SELECTED FOR SENSITIVITY TO MMS OR UV

Strain	Class[a)]	Transformation	Transfection	UV sensitivity	MMS sensitivity (m.i.c.)	MIT sensitivity (m.i.c.)	DNA uptake
PB202	Parental	1	1	1	500 μg/ml	0.1 μg/ml	1
PB1633	I	0.10	0.20	0.010	30 "	0.05 "	1
PB1634		0.30	0.15	0.012	30 "	0.05 "	1
PB1635		0.30	0.25	0.007	30 "	0.05 "	1
PB1641		0.10	0.30	0.015	50 "	0.05 "	0.8
PB1636	II	0.06	0.18	0.016	30 "	0.01 "	1
PB1639		0.01	0.05	0.033	30 "	0.01 "	1
PB1640		0.02	0.10	0.022	20 "	0.01 "	0.7
PB1642	III	0.70	0.30	0.006	300 "	0.10 "	0.7

a) Classification criteria

Class I - Reduced: transformation, transfection, resistance to UV and MMS;
normal : resistance to MIT; DNA uptake.

Class II - Reduced: transformation, transfection, resistance to UV, MMS and MIT;
normal : DNA uptake.

Class III - Reduced: transfection, resistance to UV;
slightly reduced: transformation, resistance to MMS;
normal : resistance to MIT; DNA uptake.

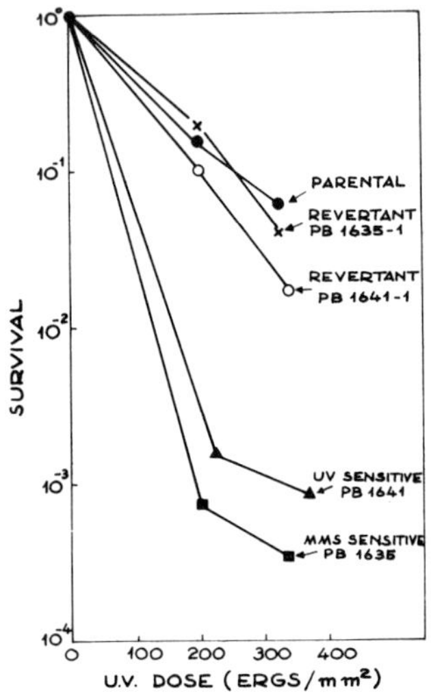

FIG. 5 - ULTRAVIOLET INACTIVATION CURVES

Competent cells of the different strains were plated on nutrient agar and exposed to varying doses of UV from a germicidal lamp; the parental strain was PB202; strain PB1641 was originally selected as UV sensitive and proved to be also MMS sensitive; strain PB1635 was isolated as MMS sensitive; strain PB1641/1 was an MMS resistant revertant of PB1641;strain PB1635/1 was an MMS resistant mutant of PB1635.

ity, as compared to the parental, with the exception of mutant PB1642 in which the activity was about one fifth of that of the wild type. These data recall those reported for a strain of B. subtilis described by Gass et al.(1971)as a polymerase-deficient mutant. These authors found also that changing the template, the deficiency in DNA polymerase activity was much more pronounced. In particular,the biggest difference was observed when activated B.subtilis DNA was used as a template.Table V shows the specific activity of DNA polymerase in crude extracts of the parental strain and of the particular mutant using different templates. When activated B. subtilis DNA was added, the activity of the

TABLE IV

DNA POLYMERASE ACTIVITY IN EXTRACTS OF STRAINS SELECTED FOR MMS OR UV SENSITIVITY

Strains	Units/mg[a)]
PB202 (parental)	0.47
PB1633	0.39
PB1634	0.40
PB1635	0.52
PB1636	0.55
PB1639	0.53
PB1640	0.46
PB1641	0.49
PB1642	0.09

a) The activity was measured according to Okazaki and Kornberg (1964) using activated calf thymus DNA as template. Activation was obtained by treating native DNA with pancreatic DNase as reported by Richardson et al. (1964).

mutant was only 5% of the control. These data have stimulated further investigation on the properties of the enzyme.

We purified DNA polymerase in parallel from the wild type and from the mutant with a new procedure.This involved precipitation with streptomycin sulfate,autolysis to destroy nucleic acids,ammonium sulfate fractionation and DEAE cellulose chromatography.The elution profile from the column is reported in Fig. 6. The upper part(A)shows the elution profiles with potassium phosphate gradients for the enzymes from the wild type and from the

TABLE V

DNA POLYMERASE ACTIVITY OF STRAIN PB1642 WITH DIFFERENT DNA TEMPLATES

Template	units/mg		%
	PB202	PB1642	of control
poly d(AT)	0.926	0.189	20.4
Activated calf thymus DNA[a)]	0.470	0.090	19
None	0.064	0.008	12
Activated _B.subtilis_ DNA	0.390	0.020	5.1

a) See note to Table IV.

mutant in two parallel columns.The parental strain gives two distinct peaks which are very reproducible in different purification and different experiments.The mutant instead shows only one peak.

In order to rule out the possibility that the two peaks of the parental were due to an artifact of the chromatographic procedure, the fractions corresponding to each peak were pooled as indicated in the figure and run again on three similar columns.The elution profiles are shown in Fig.6 (B, C, D). The peak of the mutant was eluted again at the same ionic strength. The pools corresponding to the two peaks of the parental were again eluted at two clear distinct ionic strengths. It must be pointed out, though not indicated in Fig. 6,that the portion of the profile of the mutant enzyme coming out at the same ionic strength as pool two,is only a tail of the main peak.In other experiments the fractions of the mutant enzyme, corresponding to that ionic strength,when run in a second chromatography, were eluted at the same ionic strength as the main peak of the mutant. We can con-

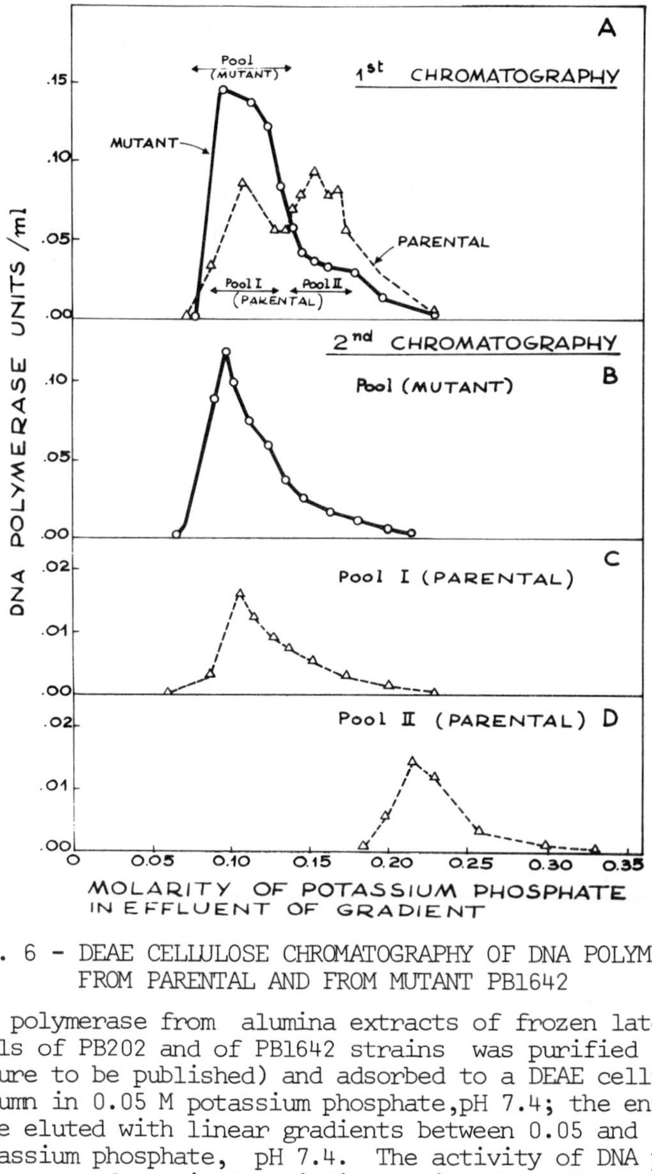

FIG. 6 - DEAE CELLULOSE CHROMATOGRAPHY OF DNA POLYMERASE FROM PARENTAL AND FROM MUTANT PB1642

DNA polymerase from alumina extracts of frozen late log cells of PB202 and of PB1642 strains was purified (procedure to be published) and adsorbed to a DEAE cellulose column in 0.05 M potassium phosphate,pH 7.4; the enzymes were eluted with linear gradients between 0.05 and 0.5 M potassium phosphate, pH 7.4. The activity of DNA polymerase was determined as indicated in Table IV.

clude therefore that with this procedure we distinguish two molecular species of DNA polymerase in the wild type and only one in the mutant, corresponding to one of those of the wild type. Recently Laipis and Ganesan (1972) have isolated a mutant with similar properties and have shown also the presence of two DNA polymerases in the wild type, one of which is missing in the mutant. It is tempting to identify the main peak of the wild type, eluted at higher ionic strength, to polymerase I, and the early peak of the wild type,together with the enzyme of the mutant, as polymerase II.Mutant PB1642 can thus be equated to the Pol A mutant of E. coli. Further enzymologic studies are in progress.

(e) Genetic studies of UV and MMS sensitive mutants

All the eight strains selected for UV or MMS sensitivity,when crossed by transformation,appeared to correspond to different mutations. Two of them,including the one deficient in DNA polymerase, have been mapped by transduction with PBS1 phage.The location on the map of the two mutations is shown in Fig. 7;they map in positions clearly different from the other UV sensitive mutants described in the literature (Hoch and Anagnostopoulos, 1970).

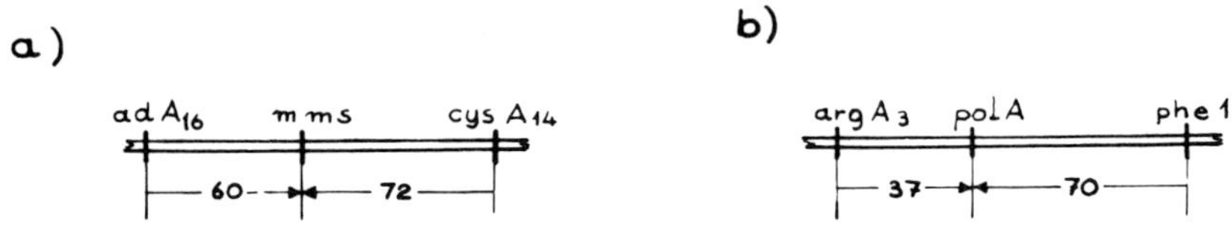

FIG. 7 - MAPPING MUTANTS PB1634 AND PB1642 BY TRANSDUCTION WITH PBS1 PHAGE

a) Location of the MMS marker of mutant PB1634; the recipient strain was PB3409 ad_{16} $cysA_{14}$ leu_8.

b) Location of Pol A mutation of strain PB1642; recipient strain was BC26 phe_{12} $argA_3$ ery(obtained from J.Copeland). The numbers represent recombination values (100 minus per cent of cotransfer) from two point crosses.

4. CONCLUSION

The isolated mutants that are most promising and worth further studies are the following: one mutant (PB1630 listed in Table I) is much more impaired in transformation than in transfection. One mutant (PB1647) is strongly temperature sensitive for transformation and transfection, whereas DNA uptake is not reduced ; the availability of a temperature sensitive mutant should help in recognizing the altered gene product. One mutant (PB1642)has properties recalling those of the Pol A mutant of *E. coli* considering its apparent deficiency in one DNA polymerase.

5. REFERENCES

ADELBERG, E. A., MANDEL, M. & CHEIN CHING CHEN, G. (1965). Biochem. Biophys. Res. Comm. 18:788.

DE LUCIA, P. & CAIRNS, J. (1969). Nature, 224:1164.

GASS, K. B., HILL, T. C., GOULIAN, M., STRAUSS, B. S. & COZZARELLI, N. R. (1971). J. Bacteriol. 108:364.

HOCH, J. A. & ANAGNOSTOPOULOS, C. (1970). J. Bacteriol. 103:295.

HOCH, J. A., BARAT, M. & ANAGNOSTOPOULOS, C. (1967). J. Bacteriol. 93:1925.

LAIPIS, F. J. & GANESAN, A. T.. J. Biochem. (in press).

OKAZAKI, T. & KORNBERG, A. (1964). J. Biol. Chem. 239:259.

OKUBO, S. & ROMIG, W. R. (1966). J. Mol. Biol. 15:440.

RICHARDSON, C . C., SCHILDKRAUT, C. L., APOSHIAN, H. V. & KORNBERG, A. (1964). J. Biol. Chem. 239:222.

RIVA, S., POLSINELLI, M. & FALASCHI, A. (1968). J. Mol. Biol. 35:347.

SEARASHI, T. & STRAUSS, B. (1965). Bioch. Biophys. Res. Comm. 20:680.

SPATZ, H. C. & TRAUTNER, T. A. (1971). Molec. Gen. Genetics, 113:174.

STEWART, C. R. (1969). J. Bacteriol. 98:1239.

YASBIN, R., WILSON, G. A. & YOUNG, F. E. (1971). Proc. Amer. Soc. Genetics, s76.

ZADRAZIL, S. & FUCIK, V. (1971). Biochem. Biophys. Res. Comm. 42:676.

HEAT SENSITIVITY OF COMPETENT AND PRE-COMPETENT CELLS OF BACILLUS SUBTILIS

LUÍS J. ARCHER

Laboratório de Microbiologia,
Instituto Gulbenkian de Ciência,
Oeiras, Portugal

ABSTRACT

Newly-transformed cultures of *B.subtilis* were fractionated in Vasurix gradients, resulting in a top fraction (T), predominantly made up of competent, pre-competent and newly-transformed cells, and a bottom fraction (B) made up almost exclusively by non-competent cells. Each fraction was heat treated at different temperatures. At various times during the heat treatment, the number of viable cells in each fraction and the number of transformants in fraction T were determined. Death curves obtained showed the presence of a shoulder and a low death rate for fraction B while the curves for the transformants showed a considerably higher death rate and no shoulder. Death curves for fraction T were decomposable in the two other curves, showing that competent and light pre-competent cells die at the same high rate as the transformants. This was confirmed by the thermal-death behaviour of the refractionation products of T fractions. When the heat treatment was performed in the absence of transforming DNA, the death curves of fraction T did not change. Heat treatment of fractions extracted at different times during competence development indicates that, before acquiring transformability, lesser dense cells are already highly heat-sensitive.

It is concluded that competent and light pre-competent cells, independently of the transformation process, are more heat sensitive, with respect to survival, than non-competent cells.

1. INTRODUCTION

It has been shown that the number of *Bacillus subtilis* transformants is reduced when, immediately before or after transformation, a competent culture is heat-treated (McCarthy and Nester, 1969). Even at temperatures at which the viability of the bulk population is not affected, the number of viable transformants decreases significantly. These findings have been interpreted as indicating the existence of some heat-labile factor required for uptake, or retention, or both of the transforming DNA (McCarthy and Nester, 1969).

More recently, McDonald (1971) has shown that *B.subtilis* transformants for streptomycin dependance, after being already phenotypically expressed, are still temperature sensitive. In this case, it is more difficult to admit that a donor DNA fragment, while already functioning in the recipient cell, could be extruded by thermal stress. No final interpretation of the results was given by the author (McDonald, 1971).

Ravin and Ma (1972) found that competent streptococci, pneumococci or *B. subtilis* cultures, when heated, respond differently to homospecific, heterospecific and hybrid DNA, and to different markers. They suggest the existence of a heat-labile, intracellular mechanism able to recognize base sequences of DNA in the course of the transformation process.

In the discussions of these papers it has been tacitly assumed that competent cells, independently of the transformation process, are as heat-sensitive as the bulk non-competent population.

This assumption is questioned in the present paper.

Competent cultures of B. subtilis usually exhibit very low transformation frequencies, making it very difficult to decide whether a decline in the number of transformants is due to a block in the transformation process, or to the absolute loss of viability of the transformable cells.

Our approach was to study the viability, upon heat-treatment, of the highly competent and of the poorly competent fractions of B. subtilis cultures fractionated in Vasurix density gradients.

The results, to be reported here, show that, independently of the transformation process, competent and light precompetent cells, as well as newly transformed cells are, in fact, considerably more heat-sensitive than non-competent cells of the same population.

2. MATERIALS AND METHODS

The bacterial strain used as recipient was B. subtilis SB 566, a tryptophan (trp_2^-)- and thymine (thy_x^-)- deficient derivative of strain 168 (Farmer and Rothman, 1965).

For development of competence, the procedure used was that of Anagnostopoulos and Spizizen (1961). Transforming DNA was isolated, from prototrophic B. subtilis 168, by the method of Saito and Miura (1963) and dissolved in 0.15 M NaCl, 0.015 M sodium citrate, pH 7.0.

The transformation reaction was carried out at 37° C with aeration, and using saturating amounts of DNA. After 20-30 min. (unless otherwise indicated) the reaction was terminated by the addition of 5 to 10 μg of deoxyribonuclase per ml. This mixture was incubated at 37° C for 2 min. Dilutions were done in Spizizen minimal medium without supplements.

Fractionation of the culture was performed by the method of Cahn and Fox (1968), except that, instead of Renografin, Vasurix 38 was used. This is a Methylglucamine Acetrizoate produced by the Laboratoires André Guerbert, France. The densities of Vasurix water dilutions were calculated by refractive index, using a Refractometer of Abbe. To correlate refractive index with density, the graph presented by Cahn and Fox (1968) for Renografin was used.

In all experiments presented in this paper, transformation, when used, was performed immediately before fractionation. The time elapsed between deoxyribonuclease addition and analysis of the fractions was about 50 min.

Thermal death experiments were performed in water-baths equilibrated at the desired temperatures. The cultures were stirred magnetically. The medium used for the heat treatment was the second transformation medium (Anagnostopoulos and Spizizen, 1961). To this medium, previously heated, aliquots of the fractionated cultures were added. By this procedure, the cultures were diluted 1/50 or 1/100. When the number of transformants was to be determined, those dilutions were 1/3 or 1/10.

Dilutions of the cultures in Spizizen minimal medium without supplements, were always plated on solid medium of the same composition and appropriatly supplemented.

3. RESULTS

(a) Heat sensitivity of fractionated cultures

Newly-transformed cultures were fractionated in Vasurix gradients, and the resulting top and bottom fractions assayed both for transformation frequencies and for viability upon heat treatment.

The fractionation was usually efficient, as estimated by the transformation frequencies of both fractions. Table I

TABLE I

TRANSFORMATION FREQUENCIES OF SAMPLES REFERRED IN THE FIGURES

Number of Fig.	Frequency of <u>trp</u>$^+$ transformants (%)			Ratio T/B
	Before fractionation	Top fraction (T)	Bottom fraction (B)	
1 A	0.5	6.7	0.006	1117
1 B	0.6	7.3	0.007	1043
1 C	0.6	8.9	0.014	635
3	0.65	12.7	0.051	249
5	12.7	12	1.4	8 (*)
6; 20 min.	0.0004	0.017	0.0007	24 (*)
6; 70 min.	0.002	0.045	0.0005	90 (*)
6;120 min.	0.1	2.4	0.01	240

(*) These low values are justified by the special conditions of the experiments to be explained later in the text.

exemplifies the values usually obtained. A total of 40 thermal death experiments were performed using a variety of temperatures ranging from 48° C to 70° C. In all experiments, the cells of the top fraction always revealed higher heat-sensitivity than those of the bottom fraction. Fig. 1 shows the general pattern of the

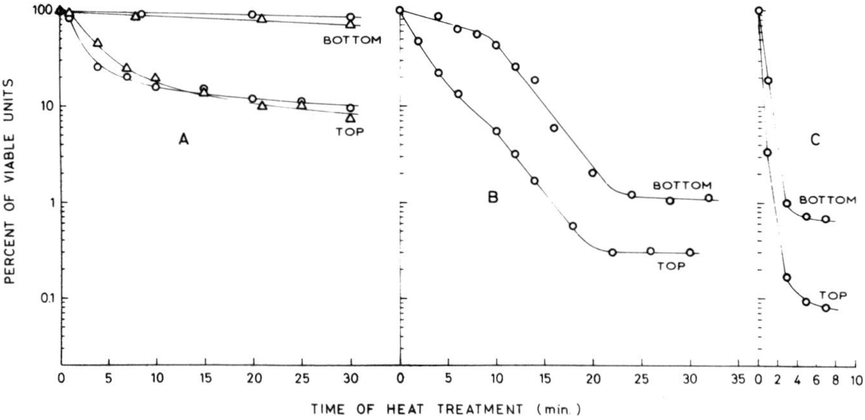

Fig. 1 - Survival curves of heat-treated fractions of competent and newly transformed cultures. Temperatures used: A = 49° C, B = 53° C; C = 57° C. O:DNA added previous to fractionation. Δ:No DNA added.

survival curves. At high temperatures (Fig. 1 C) the only detectable difference in death rates between top and bottom fraction took place in the very first moments of the heat treatment. This difference could be studied in more detail as the temperature was lowered. At temperatures around 53° C (Fig. 1 B), the survival curve of the bottom fraction showed an initial shoulder, followed by an exponential death phase, and, finally, by a stationary phase. At the same temperatures, the survival curve of the top fraction exhibited no shoulder, but had an initial convex-downward part, followed by exponential and stationary phases which run in parallel with those of the bottom fraction. At temperatures around 49° C (Fig. 1 A) the pattern of the survival curves was essentially the same as the one indicated for the 53° C experiment. Note that during the time period represented in Fig. 1 A only the first part of the death phases was observed.

The different behaviour of top and bottom fraction cells to heat, can not be attributed to the incorporation of transforming DNA, which takes place, mainly, in the top fraction cells. In fact, a series of experiments were performed in parallel with competent cultures to which transforming DNA had and had not been added, the results being essentially the same. Small deviations in one direction, found in some experiments (Fig.1 A), were reversed in other experiments, and can be better explained by random variations in the fractionation efficiency, as will be later indicated.

Since the cells of the bottom fraction are in a multiplying state, while the majority of the top fraction cells is not (Nester and Stocker, 1963), it is possible that, at low temperatures, the survival curves of the top fraction reflect thermal death only, while the bottom fraction survival curves represent the net result of death and growth. This would account for the difference between the two fraction curves. To check this possibility, the optical density of the total population was followed during heat treatment. Fig. 2 shows the results of two of these experiments. It can be seen that, during treatment at 49° C, there is in fact a significant increase in biomass simultaneously with a decrease in viability (Fig. 1). However, even if that increase in biomass represented cell multiplication, its extent would not account for the difference between the survival curves of the two fractions (compare Figs. 1 A and 2).

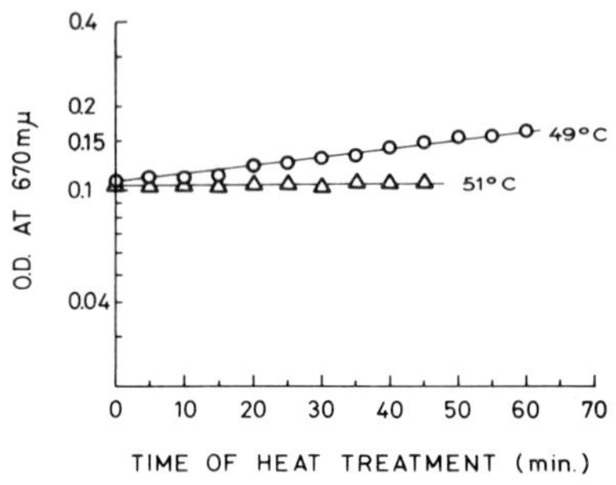

Fig. 2 - Optical densities during heat-treatments.

It may be suggested, therefore, that the different survival capacity of the two cell fractions upon being heated is a reflection of their different composition in two cell types differently affected by thermal inactivation. Competent or light cells appear to be more heat-sensitive than non-competent cells.

(b) Analysis of the top fraction curve

A series of experiments were performed in which, in addition to the survival of top and bottom fractions, the survival of the transformants from the top fraction was also observed. Fig. 3 gives the results of one of these experiments.

The data show that the heat-sensitivity of the top fraction population is intermediate between those of the bottom fraction and of the transformants. This might signify that these transformants, in addition to the high thermal-sensitivity of competent cells, also reveal the heat-labile factor needed for some step of the transformation process (McCarthy and Nester, 1969). Alternatively, the results may be interpreted in terms of contamination of the top fraction by heavy, non-competent cells.

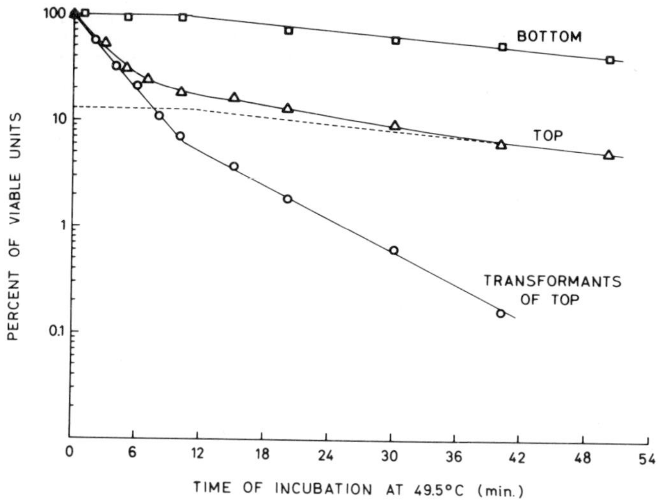

Fig. 3 - Survival curves of fractions of a newly transformed culture.Extrapolation in dashed line (see text).

All truly light cells would become as quickly non viable by heat as the newly transformed cells, which are always light (Dooley, Hadden and Nester, 1971). In contrast, contaminating heavy cells would be highly heat-resistant. Consequently the top fraction, when heated, would behave as if it were composed of a majority of fast-dying and a minority of long-lasting cells.And, in fact, this agrees with the non-exponential death observed in the top fraction up to that point when the initially minoritary cell type becomes the sole detectable element. From that point on, the top fraction is indistinguishable from the bottom fraction.

In order to obtain a quantitative test of this hypothesis, we analysed the data of a series of experiments similar to the one represented in Fig. 3.

According to Fig. 3, the top fraction behaves as bottom fraction after 40 min. of heat treatment. Extrapolating back the 40 to 50 min.straight line in parallel with the bottom fraction curve, the extrapolated line intercepts the ordinate axis at the value of 13%. This would indicate that, at time zero of the thermal-death experiment, 13% of the top fraction cells belonged to the bottom-fraction type (heavy cells), the remainder (87%) being truly light cells.

Using this information, we tried to reconstruct the survival curve of the heterogeneous top fraction in Fig. 4.

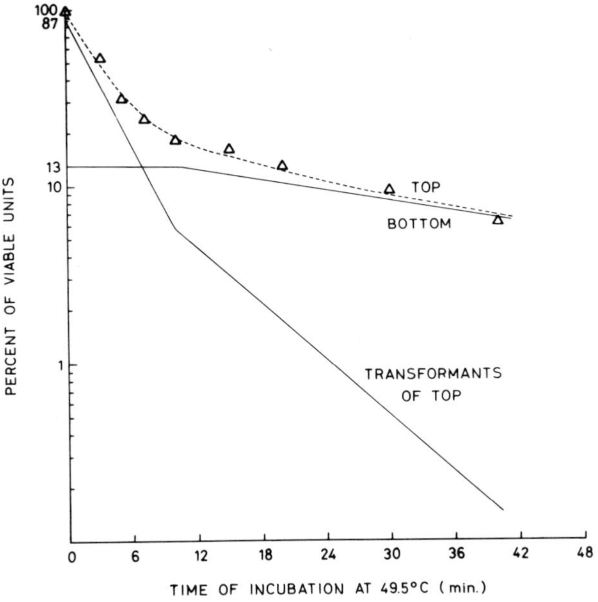

Fig. 4 - Reconstruction of the top fraction curve.
—— curves taken from Fig. 3.
---- sum of the values of the two other curves.
Δ experimental points for the top fraction (from Fig. 3).

We assumed, then, that the newly-transformed cells from the top fraction represent, as far as thermal-death rate is concerned, the 87% of truly light cells of the top fraction. Accordingly, we replotted their survival curve (from Fig. 3) adjusting its origin to 87% (Fig. 4).

Since 13% of the top fraction behaves as the bottom fraction (see above) we replotted in Fig. 4 the bottom fraction curve from Fig. 3, adjusting its origin to 13%.

Adding the values indicated by the two curves, for each time, a theoretical thermal-death curve of the total top fraction was constructed (dashed line in Fig. 4). This curve is in very good agreement with the experimental points obtained for that fraction.

This indicates that truly light cells suffer thermal-death at a rate indistinguishable from that of the transformants, and that, consequently, the survival curve of the top fraction is decomposable in the two other curves.

In order to confirm experimentally this finding, an aliquot of the same top fraction used for the experiment represented in Fig. 3 was refractionated in a Vasurix gradient, and the refractionation products subjected to identical heat treatment. The results, represented in Fig. 5, show that the initial top fraction in fact contained heat-resistant bottom-type cells. Consequently, the refractionated top fraction (Fig. 5: "top of top") followed for a longer period of time the survival curve of the transformants, as would be predicted on the basis of its higher purity.

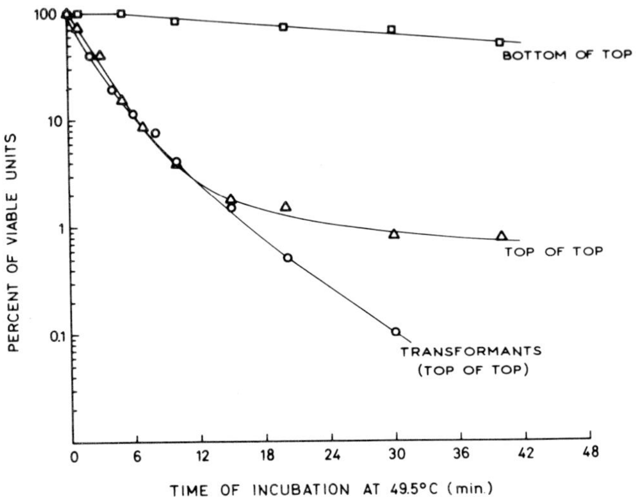

Fig. 5 - Survival curves of refractionation products of top fraction of Fig. 3.

These results show that the different thermal behaviour of top and bottom fractions does not originate in different environmental conditions of the two fractions(as different Vasurix concentrations)but rather shows that the high heat sensitivity of competent or light cells is indistinguishable from that of the transformants analysed.

(c) Heat-sensitivity of pre-competent cells

A further question refers to when does the heat-sensitivity of light cells develop. Are pre-competent cells heat sensitive?

It is known that pre-competent cells, long before becoming transformable,acquire the low buoyant density of the competent cells,and band, therefore, in the top fraction of the gradient (Dooley, et al., 1971).

If heat sensitivity developed in pace with transformability, a correlation between the two would be found in the top fractions of different experiments. Table II shows that this is

TABLE II

ANALYSIS OF THE FRACTIONS OF NEWLY-TRANSFORMED CULTURES

Experiment number	Bottom fraction	Top fraction		
	Frequency of trp$^+$ transformants (%)		Competent cells (%)(1)	Heat sensit. cells (%)(2)
1	0.05	12.7	178	87
1 refr.*	1.4	12.0	111	98.4
2	0.003	1.54	-	87
2 refr.*	0.16	2.9	-	98
3	0.004	7	-	96
3 refr.*	0.24	14	-	98.6
4	0.0006	0.7	11	60
5	0.005	2.4	33.6	74
6	0.03	2.6	31	89
7	0.007	7.3	112	94
8	0.00006	2.2	12.5	99.5
9	0.003	3.4	-	97.5

* Refractionation of the top fraction of the previous experiment.

(1) Calculated by cotransfer of unliked trp and thy markers (Goodgal, 1961).

(2) Calculated from thermal death curves as indicated in Fig. 3 and in text.

not the case. The presented discrepancy can be best understood by admitting that light pre-competent cells are about as heat-sensitive as competent cells and that the ratio between these two cell types changes from experiment to experiment.

A confirmation of this interpretation was found in a series of experiments in which fractionation and analysis of its products were performed at different times during development of competence. At 10, 60, and 110 min. after the 1/10 dilution of the method of competence development (Anagnostopoulos and Spizizen, 1961), a sample was withdrawn and incubated for 10 min. in the presence of transforming DNA. The reaction was terminated by 2 min. incubation with deoxyribonuclease. Fractionation in Vasurix gradients followed. The fractions were immediately analysed as described above. Table III and Fig. 6 show results of one of these experiments.

TABLE III

ANALYSIS OF TOP FRACTIONS EXTRACTED AT DIFFERENT TIMES DURING DEVELOPMENT OF COMPETENCE

Sample	Percent of cells (1)	Frequency of $\underline{trp}^+$ transformants (%)	Competent cells (%) (2)	Heat-sensit. cells (%) (3)
20 min.	21	0.017	0.051	76
70 min.	50	0.045	1.1	78
120 min.	6	2.4	36	82

(1) Estimated as $\frac{T}{T+B} \times 100$, where T = number of cells in the top fraction, and B = number of cells in the bottom fraction.

(2) Calculated as indicated in Table II.

(3) Calculated from thermal-death curves (see Fig. 6).

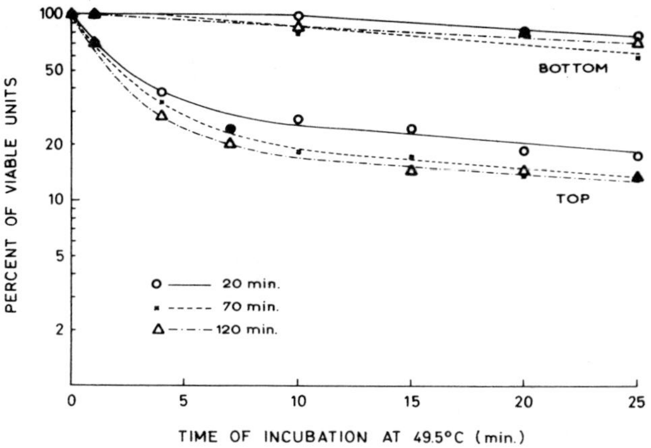

Fig. 6 - Survival curves of fractions extracted at different times from a competence developing culture.

At the three times considered, the top fractions, in spite of having drastically different transformation frequencies, showed comparable contents of heat-sensitive cells. This indicates that, previously to the acquisition of transformability, light cells are already as heat-sensitivity as competent cells.

It is known that the highest proportion of light cells in a competence-developing culture occurs about 45 min. before the peak of competence,when the number of transformants is still very low (Dooley, et al., 1971).

This was also the case in our experiments as indicated for the 70 min. sample of Table III. Since the fractionation for that sample was about as efficient as for the others, this peak of light cells corresponds also to a peak of heat sensitive cells in the total culture. Other experiments (results not given) confirm this result. It seems, therefore, that heat-sensitivity affects all the light cells - competent and pre-competent. In addition, we may conclude that thermal death curves seem to be a better measure of the efficiency of fractionation than the frequency of transformation or the calculated percentage of competent cells.

(d) Outgrowth from the heat-sensitivity

The bottom fractions obtained from refractionation of top fractions ("bottom of top") always accused a transformation frequency considerably higher than that of the bottom fraction of the first fractionation (Table II).

Considering that refractionation was completed about 100 min. after the termination of the cells-DNA reaction, a possible interpretation is that, by that time, some transformants already acquired the higher density of the non-competent cells.

The heat-sensitivity of the transformants from these "bottom of top" fractions was not determined. However, if they still possessed their previous high heat-sensitivity,their presence would probably not pass unnoticed in the thermal-death curve of the bottom fraction(compare with top fractions of Table II). This, however, was not the case (Fig. 5).

The heat survival curves of the transformants always showed two distinct slopes (see, as examples, Fig. 3 and 5). The reason for this is unknown. A possible interpretation, however, is that the second slope represents transformants starting recovery from heat-sensitivity.

4. DISCUSSION

The results presented in this paper are not in contradiction with the existence, for some period of time, of a heat-labile factor,specifically needed for the transformation process (McCarthy and Nester, 1969; Ravin and Ma, 1972). They only add that,even independently of that process,competent and light pre-competent cells suffer thermal-death at a rate which is significantly higher than that of the non-competent cells of the same population. Consequently, this finding warns us that the difference between thermal-death rates of transformants and of total viable cells, should not be considered a true measure of the impairement of competence or of the transformation process.

The fact that the top fraction survival curve is decomposable in the two other curves (Figs. 3 and 4), and the refractionation experiments, show that light pre-competent and competent cells have a thermal-death rate undistinguishable from that of the transformants assayed. Considering the conditions of the experiments, we may conclude that 50 min. after termination of the transformation reaction, no heat-labile factor existed, specifically needed for the completion or expression of transformation. Thermal-death of these transformants was independent from the transformation process.

This may also be the case in the transformants found by McDonald (1971) as heat-sensitive, 3 to 6 hours after DNAse addition. If this proves true, and knowing the conditions of McDonald's experiments, it would mean that these transformants recovered normal heat-resistance rather late,even after phenotypic expression, and shortly before they start dividing.

The primary target of heat is not yet known. There are indications, however, that ribosomal RNA synthesis is needed for recovery from thermal injury (Tomlins and Ordal, 1971) and that cell envelope and its permeability are affected by heat (Nash and Sinclair, 1968).

It is also well known that competent cells do not synthesize ribosomal RNA (McCarthy and Nester, 1967), accumulate a high number of unusually large mesosomes (Wolstenholme,Vermeulen and Venema, 1966), and have altered cell envelope and permeability (Young, Spizizen and Crawford, 1963; Miller, Zsigray and Landman, 1967).

It would not be surprising, therefore,if some of these properties of the competent cells were related with their high heat-sensitivity.

Since the state of competence is quite well characterized, and five different steps were already distinguished in its development (Dooley, et al., 1971), the study of the heat sensitivity along these steps may prove useful for the purpose of pinpointing the mode of action of heat in thermal cell death.

ACKNOWLEDGEMENTS

Mrs. Maria Cândida Lopes and Mrs. Margarida do Vale Passos performed most of the experimental work.

5. REFERENCES

ANAGNOSTOPOULOS, C., & SPIZIZEN, J. (1961). J. Bacteriol. 81:741.

CAHN, F. H., & FOX, M. S. (1968). J. Bacteriol. 95:867.

DOOLEY, D. C., HADDEN, C. T., & NESTER, E. (1971). J. Bacteriol. 108:668.

FARMER, J. L., & ROTHMAN, F. (1965). J. Bacteriol. 89:262.

GOODGAL, S. H. (1961). J. Gen. Physiol. 45:205.

MCCARTHY, C., & NESTER, E. W. (1967). J. Bacteriol. 94:131.

MCCARTHY, C., & NESTER, E. W. (1969). J. Bacteriol. 97:162.

MCDONALD, W. C. (1971). Can. J. Microbiol. 17:823.

MILLER, I. L., ZSIGRAY, R. M., & LANDMAN, O. E. (1967). J. Gen. Microbiol. 49:513.

NASH, C. H., & SINCLAIR, N. A. (1968). Can. J. Microbiol. 14:691.

NESTER, E. W., & STOCKER, B. A. D. (1963). J. Bacteriol. 86:785.

RAVIN, A. W., & MA, M. (1972). J. Bacteriol. 109:616.

SAITO, H., & MIURA, K. (1963). Biochim. Biophys. Acta 72:619.

TOMLINS, R. I., & ORDAL, J. (1971). J. Bacteriol. 105:512.

WOLSTENHOLME, D. R., VERMEULEN, C. A., & VENEMA, G. (1966). J. Bacteriol. 92:1111.

YOUNG, F. E., SPIZIZEN, J., & CRAWFORD, I. P. (1963). J. Biol. Chem. 238:3119.

AN APPROACH TO THE STUDY OF COMPETENCE REGULATION IN *B. SUBTILIS* GROWING IN A CHEMOSTAT

A. PORTOLÉS
R. LÓPEZ
A. TAPIA

Institute "Jaime Ferrán" of Microbiology.C.S.I.C.
Madrid, Spain

ABSTRACT

Development of competence in *B. subtilis* strains growing in a chemostat has been studied by transformation and transfection. Maximal transformation frequencies were detected at doubling times of 2.5 hrs. and 6.5 hrs. Stimulatory (e.g. arginine, histidine) and inhibitory (e.g. glutamic acid, alanine) amino acids showed reverse behavior patterns with respect to competence development at the above doubling time. Kinetic evidence demonstrates that at Dt = 6.5 hrs. these amino acids show an effect similar to that described in batch cultures by Bott & Wilson, while at Dt = 2.5 hrs. the effect was the opposite. Based on these and some previous observations we suggest that there are at least two different ways to get the competent state which, in some respects has a relationship with the dilution rate where the inhibitory and stimulatory amino acids seem to play regulatory roles.

1. INTRODUCTION

The remarkable importance of regulation in biological phenomena is well-known. For this reason special attention has been paid to the regulation of enzyme synthesis and activities. The most recent papers on transformation demonstrate the tendency of many workers to study the regulation of competence. In fact, a considerable effort has been made in order to simplify the procedures for studying these regulative processes in transformation.

Nevertheless, as Clark & Lilly (1969) have pointed out, little attention has been devoted to continuous culture as a tool for investigating regulation phenomena. They reported that " In batch cultures the chemical environment is continually changing thereby introducing another varying parameter into what is already a complex situation, and it is difficult to vary the growth rate except by altering the carbon and/or energy source or by changing the temperature. In a chemostat the growth rates of cultures can be limited by any of the nutrients essential for growth under exactly defined conditions".

These facts were our main reason for undertaking a research program on the regulation of competence in B. subtilis growing in a chemostat, and for concentrating particularly on their variations as a function of both the pre-sporulation state and the influence that certain amino acids in the medium have on competence at different doubling times.

With regard to regulation studies, the classical method to get the competent state "in vitro" using the step-down procedure (Anagnostopoulos & Spizizen, 1961) leads to numerous difficulties when used either in a chemostat or in batch cultures, since, as Bott & Wilson (1967) have reported, it can induce false results as a consequence of the 4 hrs plus 90 minutes regimen. In this sense, the modifications they introduced in single growth medium cultures have been a basic aid in studying the problem of competent state regulation.

Even with these conditions, however, we cannot avoid the biological problems of exhausting the medium and accumulating catabolites (Tomasz and Mosser, 1966; Horvath, 1967).We have previously checked some variations in the competence wave as a function of the concentration of a limiting nutrient (López *et al*. 1970).The above remarks lead us to think that the continuous culture could be one of the most important tools to elucidate certain problems involving competence regulation, since it provides us with bacteria cells whose physiological conditions are lasting, homogenous and reproducible. Also, an equilibrium between anabolites and catabolites can be obtained with this system.

2. MATERIAL AND METHODS

a) Bacterial strains

B.subtilis SB-25 (his_2^- trp_2^-) was used as the recipient strain for transformation experiments. *B.subtilis* AR1T, a prototrophic strain obtained in our laboratory for transformation from *B. subtilis* SB-25, was used as a recipient for transfection experiments. *B. subtilis* MC-1 (rec 1^+, rec 2^-, her^+) was used as an indicator strain in transfection.

b) Preparation of DNA

The bacterial donor DNA was obtained at the end of the exponential phase from *B. subtilis* WT grown in Penassay medium (Difco) following a modification of Kirby's method(Venema,*et al*., 1965).The DNA from the phage SPP1 was obtained and purified in a

Cs Cl gradient by the method previously described by Rottländer & Trautner (1970).

c) Growth in continuous culture

The competent recipient culture was grown in a chemostat in similar conditions to that previously described (López *et al.*, 1970), unless otherwise stated.

d) Growth media

For the transformation experiments, the bacteria were grown in a chemostat using the Bott & Wilson (1968) medium where the required amino acids were added to a concentration of 50 µg/ml. For the transfection experiments, the prototrophic AR1T strain was grown in a slightly modified Bott & Wilson medium as we will describe in our results. TY medium (Spatz & Trautner, 1970) was used to grow the indicator strain in the usual way.

e) Transformation and transfection

Transformation was made as we have previously described (López *et al.*, 1970). The transfection experiments were carried out as follows: 0.8 ml of competent cells were incubated with 0.2 ml of phage DNA (20 µg) at 37° C as described by Spatz & Trautner (1970).

f) Estimation of the number of refractile spores

The incidence of refractile spores as a fraction of the total population was determined under the phase-contrast microscope as described by Dawes and Thornley (1970).

3. RESULTS

Eichhorn and co-workers (1968) using a similar medium to that described by Anagnostopoulos & Spizizen (1961),as we ourselves just discovered almost simultaneously using a Bott & Wilson medium, encountered the possibility of getting and keeping a lasting state of competence in a chemostat.Eichhorn *et al*.(1968) use a single doubling time in their experiments while we have studied the variations of the competent state as a function of the doubling time.

We ought to point out that in our early experiments of transformation we have used a double auxotrophic strain for two closely linked markers in order to decrease the frequency of retromutation in continuous culture (Novick & Szilard,1950).We are currently using transfection since this procedure has some advantages:We can obtain higher levels in the expression of competence in a shorter time and, in addition, we can vary at will in the medium the concentration of the amino acid whose metabolic role we are studying.

Our early results show a maximum of competence at a doubling time of 2.5 hrs.On the other hand, a close relationship between competent state, sporulation and an intact TCA cycle has been reported in *B.subtilis* by several authors (Spizizen *et al*., 1963).

However, it was difficult to correlate these conclusions with our results because at a doubling time as low as 2.5 hrs., the rate of pre-sporulation ought to be negligible. In fact, Dawes and Thornley (1970) found a relationship between pre-sporulation and doubling time similar to that shown in Fig. 1.

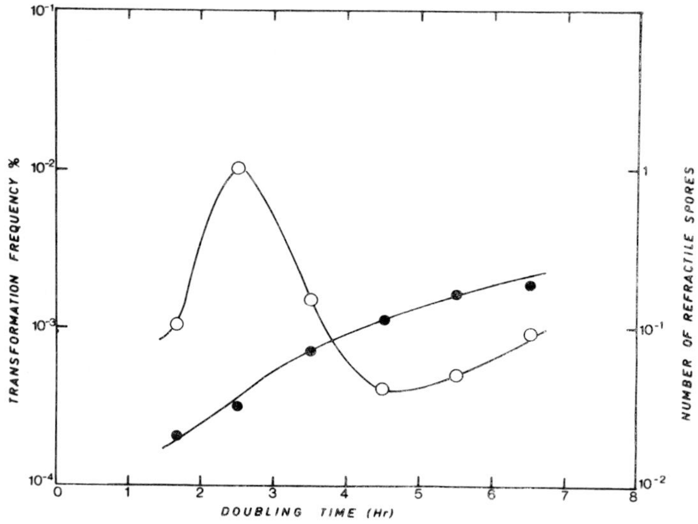

Fig. 1 - Relationship between doubling time, transformation frequency (O) and pre-sporulation (●). Each point is the average value of three experiments made every 24 hrs.

It is evident that a maximum of competence (Dt=2.5 hrs) appears when the number of pre-spores in the chemostat is remarkably low and consequently the possibility of finding TCA cycles intact is also very low.

In order to clarify these contradictory results we performed some experiments to study the influence as well as the possible regulatory effects that the amino acids, qualified as stimulatory or inhibitory by Bott & Wilson (1967), could have on the development of competence in cultures growing at different doubling times.

Fig. 2 shows the variations of the frequency of transfection at different doubling times when the recipient bacteria were grown in a Bott and Wilson medium. Once again, transfection experiments confirm the results obtained with transformation. Two maximal values of competence appear at Dt = 2.5 hrs and Dt = 6.5hrs,the former being the highest one in our experimental conditions (Transfection frequency from 3 x 10^{-2} to 1 x 10^{-2}).

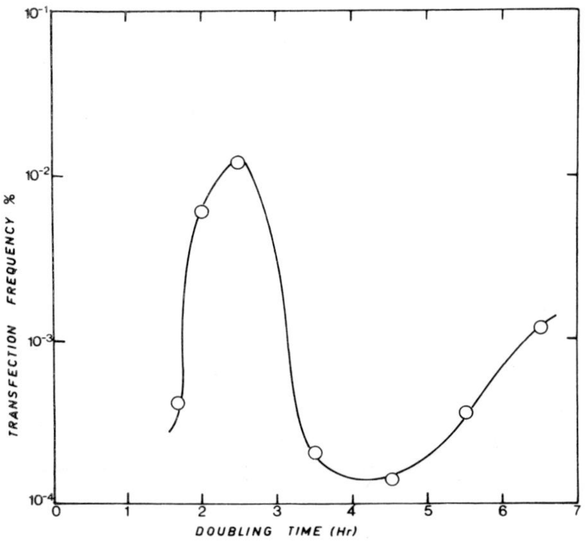

Fig. 2 - Relationship between transfection frequency and doubling time when the recipient cells were grown in a basic Bott & Wilson medium.

Figure 3 illustrates the results obtained in transfection frequency, when the amino acids were withdrawn from the basic Bott and Wilson medium. These experiments reveal the existence of a single maximum of competence at a doubling time of

2.5hrs while the other maximum at 6.5hrs completely disappeared. These results are a new clue for thinking that there are at least two different ways to get the competent state.

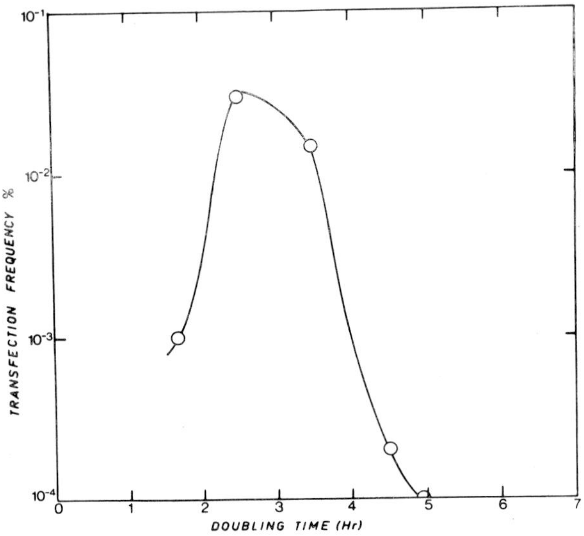

Fig. 3 - Relationship between transfection frequency and doubling time of the cells growing in a minimal salts medium.

In order to characterize the behaviour of individual amino acids considered as inhibitory (e.g.glutamic acid,alanine), and stimulatory (e.g. arginine, histidine) in batch cultures, we performed a new experiment in which the amino acid under study was kept at a constant concentration throughout the whole experiment.

The data in Fig. 4 show that a constant concentration of either 50 µg/ml of glutamic-acid or 50 µg/ml of alanine increases the frequency of transfection for doubling times ranging from 1.7 hrs to 5.5 hrs with respect to the basic curve, while the level of competence dropped at higher doubling times.

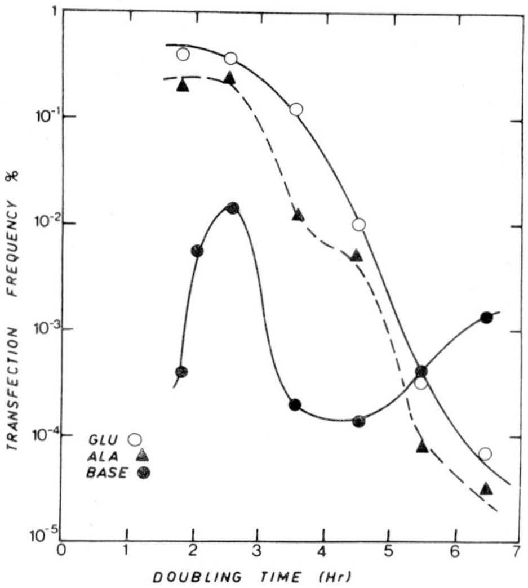

Fig. 4 - Effect of a constant concentration (50 µg/ml) of L-glutamic acid (O) and L-alanine (▲) on transfection as a function of the doubling time. (●) Transfection in Bott & Wilson basic medium (Base).

In addition, we carried out similar experiments using a constant concentration of the stimulatory amino acids.Arginine and histidine decreased the level of competence with reference to the basic curve,in a range from 1.7 hrs to 3.3 hrs while they increased the transfection from 3.4 hrs to 6.5 hrs as is shown in Fig. 5.

The above data suggest the strong influence that the doubling time has on the development of competence. Thus, we focused our attention on the doubling times of 2.5 hrs and 6.5 hrs,each of which produced a clearly distinct behaviour pattern. In this way, we performed several experiments to study the kinetics of transfection when the recipient culture was pulsed with variable amounts of amino acids.The pulsed experiments pre-

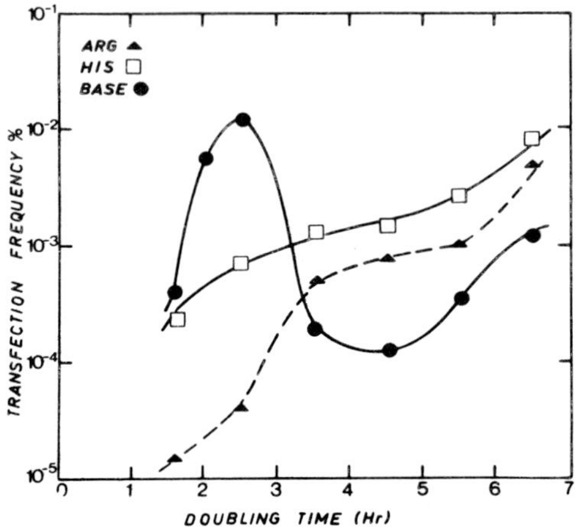

Fig. 5 - Effect of a constant concentration 50 μg/ml of L-arginine (▲) and L-histidine (□) on transfection as a function of the doubling time. (●): Transfection in a Bott & Wilson base medium. The concentration of the amino acids studied was kept constant as described in Fig. 4.

sented in Fig. 6 provided further evidence of the previous conclusion, since arginine inhibits competence more than two log. units when the amino acid concentration was increased from 0 to 100 μg/ml in the chemostat where the recipient bacteria were growing at a constant doubling time of 2.5 hrs. In contrast, when glutamic acid was added to the salts minimal medium, the competence was kept constant or slightly stimulated.

Similar experiments were carried out at a doubling time of 6.5 hrs by using the basic Bott & Wilson medium. We can see from the results shown in Fig. 7 that glutamic acid inhibits completely the development of competence while arginine stimulates it in a moderate way up to a concentration of 15 μg/ml, after which time competence remains constant.

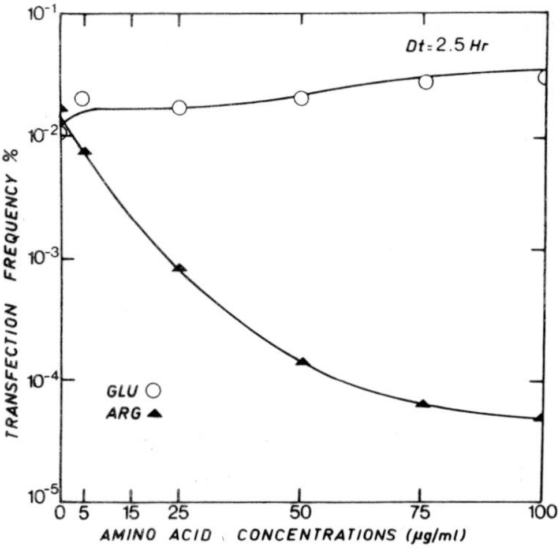

Fig. 6 - Kinetics of transfection at a constant doubling time (2.5 hrs) in the presence of different amounts of L-glutamic acid (○) or L-arginine (▲).
The final concentrations in the chemostat were calculated as described in Fig. 4 and added to the minimal salts medium. When the amino-acid concentration was modified, an interval of 48 hrs was observed between one experiment and the other in order to preserve the steady-state of the culture.

The results indicated in Figs. 6 and 7 emphasize the fact that the same amino acid (either glutamic acid or arginine) shows absolutely contrary experimental patterns for the two different doubling times studied.

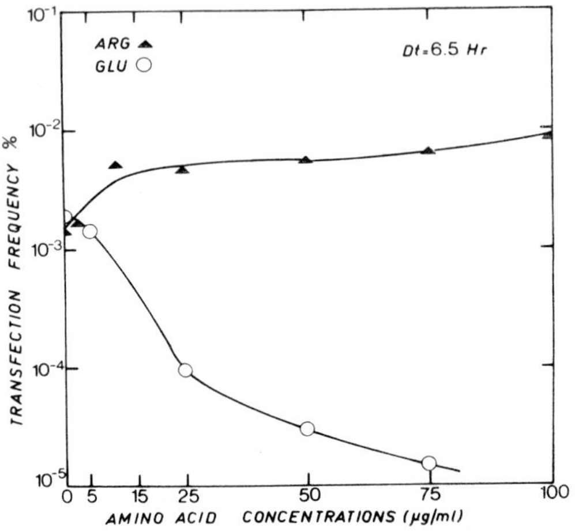

Fig. 7 - Kinetics of transfection at a constant doubling time (6.5 hrs) in the presence of different amounts of L-glutamic acid (○) or L-arginine (▲).

4. DISCUSSION

The above results demonstrate the efficiency of using the continuous culture to study transfection and transformation in B.subtilis growing in the defined minimal medium described by Bott & Wilson. These authors have pointed out (1968) that the competent state is associated with a specific phase of growth, and an altered growth rate in the logarithmic phase has little effect on competence development.

Our results(Fig. 1)indicate that in continuous culture the development of competence, strongly influenced by the growth rate, shows maximal frequencies at 2.5 hr and 6.5 hr doubling times in our experimental conditions.

On the other hand, Bott & Wilson (1967) have reported that in the absence of stimulatory amino acids (minimal salts medium), the maximal frequency of competence decreases up to 30% with respect to that obtained in the supplemented medium. Nevertheless, we observe (Fig. 3) that at a doubling time of 2.5 hrs, the maximal competence remained unaltered or was slightly stimulated when the recipient bacteria were grown in the minimal salts medium,while at doubling times up to 4.5 hrs,the competent state disappears completely.

This result could suggest that when the bacteria are growing in a continuous culture at a doubling time of 2.5 hrs, the physiological conditions that lead to competence differ markedly from those that lead to competence in batch cultures. This conclusion is reenforced by the fact that pre-sporulation is very low at the above doubling time (Fig. 1).This low pre-sporulation rate, when competence is very high, does not agree with the suggestion made by several authors (Spizizen *et al.*, 1966) that pre-sporulation has a positive effect on the development of competence.

Moreover, when these experiments were performed in the presence of 50 μg/ml of either L-glutamic acid or L-alanine (inhibitory amino acids)in the chemostat as shown in Fig. 4,a sharp increase in the competence level was detected at a doubling time of 2.5 hrs while at 6.5 hrs, the competent state was clearly inhibited.

If the competence curves were made in the presence of 50 μg/ml of either arginine or histidine (stimulatory amino acids) we got a reverse phenomenon (Fig. 5). This is to say, at a doubling time of 2.5 hrs, the level of competence was strongly inhibited, but it was stimulated when the bacteria were growing at a doubling time of 6.5 hrs.

This seems to demonstrate clearly that the competence phenomenon is wholly independent from the TCA cycle since L-glutamic acid and L-alanine are specific inhibitors of aconitase (Hanson and Cox, 1967). Furthermore the fact that arginine stimulates competence at a doubling time of 6.5 hrs completely contradicts Felkner's assumption (1970) that an intact TCA cycle is necessary for the achievement of competence. These results, as can be seen, indicate that competence is more decidedly influenced by the doubling time than by an intact TCA cycle.

In this situation when the concentration of certain amino acids (e.g. glutamic acid) are modified, competence gradually disappears, but it later regains the final value described in this work. This value is thereafter kept constant when the steady state is reached. The above leads us to believe that several amino acids may control the development of competence by repression or derepression phenomena.

From the data shown we can conclude that the ways to get competence at doubling times of 2.5 hrs and 6.5 hrs are completely different. The manner of achieving the competent state at a doubling time of 6.5 hrs is similar to the conditions for reaching competence in batch cultures. This assumption is supported by the effect of certain amino acids on the final expression of competence.

These conclusions could be related to those pointed out by Ephrati-Elizur (1968) in batch cultures, who suggested that competence in the early log phase should differ in some important aspects from competence in the early stationary phase.

Furthermore, the differing resistance against certain drugs (e.g. penicillin) of the competent population with respect to the total population reported by Nester (1964) in B. subtilis, more recently by Pakula and Spencer (1972) in Streptococcus in batch cultures and checked by ourselves in continuous cultures (López et al., 1972)at doubling times as low as 1.7 hrs, 2.5 hrs and 3.5 hrs might not be due in our cases to either a lack of cell wall synthesis or of DNA replication.

Finally, our studies suggest some modifications to the model proposed by Erickson(1970)in the B.subtilis transformation system. Whereas he rules out the evolutionary role of transformation in this microorganism, we have found that the recipient cells actively growing in a chemostat can take up DNA and are capable of expressing genetic markers.

ACKNOWLEDGEMENTS

We would like to thank Prof. A.T.Trautner who accepted one of us (A.T.) as a guest at his laboratory and kindly gave us B. subtilis strain MC1 and the SPP1 phage.

5. REFERENCES

ANAGNOSTOPOULOS, C. & SPIZIZEN, J. (1961). J. Bacteriol. 81:741.

BOTT, K. F. & WILSON, G. A. (1967). J. Bacteriol. 94:562.

BOTT, K. F. & WILSON, G. A. (1968). Bacteriol. Rev. 32:370.

CLARK, P. H. & LILLY, M. D. (1969). 19th. Symp. Soc. for gen. Microbiol. (London) At the Univ. Press pag. 113

DAWES, I. W. & THORNLEY, J. H. M. (1970). J. gen. Microbiol. 62:49.

EICHHORN, H. H., ARIKAWA, S. & ZAMENHOF, S. (1968). Can. J. Microbiol. 14:290.

EPHRATI-ELIZUR, E. (1968). Genet. Res. 11:83.

ERICKSON, R. J. (1970). Curr. Topics. Microbiol. and Immunol. 53:149.

FELKNER, I. C. & WYSS, O. (1970). Biochem. Biophy. Res. Comm. 41:901.

HANSON, R. S. & COX, D. P. (1967). J. Bacteriol. 93:1777.

HERBERT, D., ELSWORTH, R. & TELLING, R. C. (1956). J. gen. Microbiol. 14:601.

HORVATH, S. (1967). J. gen. Microbiol. 48:215.

LÓPEZ, R., PORTOLÉS, A. & PÉREZ UREÑA, M. T. (1970). Life Sciences 9:1141.

LÓPEZ, R., PÉREZ UREÑA, M. T., ESPINOSA, M. & PORTOLÉS, A. (1972 Arch. Mikrobiol. 82:206.

NESTER, E. W. (1964). J. Bacteriol. 87:131.

NOVICK, A. & SZILARD, I. (1950). Proc. Natl. Acad. Sci.(Washingt 36:708.

PAKULA, R. & SPENCER, L. R. (1972). Can. J. Microbiol. 18:487.

ROTTLÄNDER, E. & TRAUTNER, T. A. (1970). Mol. Gen. Genet. 108:47

SPATZ, H. CH. & TRAUTNER, T. A. (1970). Mol. Gen. Genet. 109:84.

SPIZIZEN, J., REILLY, B. & DAHL, B. (1963). Proc. Intern. Congr. Genetics 11th, The Hague, 1:31.

SPIZIZEN, J., REILLY, B. E. & EVANS, A. H. (1966). Ann. Rev. Microbiol. 20:371.

TOMASZ, A. & MOSSER, J. L. (1966). Proc. Nat. Acad. Sci. (Washington) 55:58.

VENEMA, G., PRITCHARD, R. H. & VENEMA SCHRODER, T. (1965). J. Bacteriol. 89:1250.

THE BINDING OF POLYDEOXYNUCLEOTIDES TO THE SURFACE OF COMPETENT PNEUMOCOCCI

ALEXANDER TOMASZ

The Rockefeller University, New York, N.Y. USA

1. INTRODUCTION

The ability of transformable pneumococci to react with DNA molecules is dependent on a complex set of alterations of the cell surface (Tomasz, 1969). Several specific macromolecular factors are known to participate in these surface alterations: a species specific, cationic "activator" substance (Tomasz and Mosser, 1966);a "receptor" for the activator(Ziegler and Tomasz, 1970);a new antigenic determinant (Nava, Galis and Beiser, 1963; Tomasz and Beiser, 1965); a surface "agglutinin" (Tomasz and Zanati, 1971). There are also lines of evidence suggesting the involvement of the cell wall growth zone (Tomasz, Zanati and Ziegler, 1971).

On the other hand, practically nothing is known about the biochemical nature of the DNA adsorbing cell surface component and about the mechanism by which DNA is transported through the surface lamellae of the recipient bacterium. In an attempt to test the specificity of the DNA binding- and transporting system we determined the ability of competent pneumococci to react with polynucleotides of different base composition and secondary structure.

2. MATERIALS AND METHODS

The following polymers were purchased from the Research Division of Collaborative Research Inc.:polydeoxyThymidylic acid (pdT), polydeoxyAdenylic acid (pdA), polydeoxyGuanylic (pdG) and Cytidylic (pdC) acids, oligodeoxyT:polyriboAdenylic acid (oligo dT:polyribo A), oligodeoxyT:oligodeoxyA(oligo dT: oligo dA).Poly dAT was obtained from the Biopolymers Division of General Biochemicals. The molecular weights of all the polymers were in the 10^6 dalton range determined by sucrose density gradient centrifugation (information provided by the manufacturers).Radioactive polydAT (^{3}H)was prepared by enzymatic synthesis from deoxyadenosine triphosphate and tritium labeled deoxythymidine triphosphate using the *E. coli* DNA polymerase (kindly supplied by Dr. Robert Klett of this University).

The final product had a specific radioactivity of 1.82×10^7 dpm per μg and showed heterogeneous molecular weight distribution in sucrose gradients (range: 10^5 to 10^6 daltons). Radioactive (^{3}H-TdR) pneumococcal DNA (specific radioactivity: 1.9×10^6 dpm per μg) was prepared by biosynthetic labeling from a streptomycin resistant mutant strain which also required thymidine for growth. Radioactive T4 DNA was prepared by adding the labeled ^{3}H-TdR to the culture medium 15 minutes after infection of the host bacteria with 5 Pfu phage per cell (specific radioactivity of DNA: 2.5×10^5 dpm per μg).Parental(single stranded, circular) and replicative form (RF-1, double stranded, circular) DNA of bacteriophage f-1 was obtained from Dr. Ken Horiuchi of this University. ^{3}H-TdR labeled DNA of adenovirus type 2 (spec. radioactivity: 6.1×10^4 dpm per μg) was provided by Dr. Walther Doerfler (Rockefeller University).

Pneumococci in the "incompetent"and"competent" physiological states were prepared by published procedures (Tomasz, 1966; Tomasz, 1970). For the determination of the effects of various polynucleotides on genetic transformation the competing polynucleotides were added to the competent cells simultaneously with the transforming homologous DNA which was used at saturating concentrations. Binding assays with radioactive polynucleotides were performed as follows : competent bacteria, stored frozen at -70°C in 10% glycerol, at cell concentrations of 1-5 x 10^7 cells per ml, were melted rapidly at 30°C. After centrifugation in the cold the cells were resuspended in fresh, cold medium at a concentration of 5 x 10^7 cells per ml. Incubation with the radioactive polynucleotides was carried out at 30°C for 10 to 20 minutes.After chilling, the bacteria were centrifuged (10000 x g 5 minutes), washed 3 times in the centrifuge with 5 ml batches of cold medium and resuspended in 2 ml medium. One ml was added to 9 ml of cold, 10% TCA and after transfer to GFA filters was used for the determination of total cell-associated DNA. The other 1 ml was first treated with pancreatic DNase (10 μg per ml) for 10 minutes at 37°C,was washed 2 times in the centrifuge,resuspended in 1 ml medium and after precipitation with cold, 10% TCA was used for the determination of nuclease resistant,cell-associated DNA. Measurement of radioactivity was done in a scintillation spectrometer.

3. RESULTS

Two types of assays were used to evaluate the ability of competent cells to "bind" heterologous polynucleotides : 1) polynucleotides were tested as potential competitive inhibitors of genetic transformation; 2) the cellular binding of the radioactive polynucleotide was tested directly.

The results are summarized briefly in the three Tables.

Table I shows that at the concentrations used all the double stranded polynucleotides tested have lowered the frequencies of transformation by homologous DNA. Interestingly, circular double stranded DNA and the "hair-pin"structured pdAT molecules were also active. On the other hand, single stranded circular (or linear) molecules as well as hybrid deoxyribo:ribo polymers were inactive.

TABLE I

INHIBITION OF TRANSFORMATION BY COMPETING POLYNUCLEOTIDES

Transforming DNA	Competing Polynucleotide (µg/ml)		Effect on Transformation: Surviving Transformants (in % of control values)
0.27 µ per ml	None		(100)
throughout	Homologous DNA	0.32	43
	Adenovirus 2 DNA	0.55	27
	Coli Phage T4 DNA	0.25	14
	Coli Phage F1, ss circ.	1.0	100
	ds circ.	1.0	10
	Poly dA:dT	7.65	10
		0.765	69
	dG:dC	100.0	12
	Oligo dT:pdA	200.0	18
	Oligo dT:p-riboA	200.0	100
	Poly dAT	100.0	2.5
		40	19.5
		20	40.3
		10	100

Table II shows that at saturating concentrations heterologous and homologous DNA-s were all bound to roughly the same extent by the bacteria.While the binding tests were clearly positive for pdAT the amount of these molecules bound by the cells was relatively low.

TABLE II

BINDING OF POLYNUCLEOTIDES TO COMPETENT PNEUMOCOCCI

Polynucleotide	Conc. per ml	Amount Bound (γ) per 10^8 Competent Cells
Pneumo DNA	3.82	0.046
	0.80	0.023
Adeno Virus 2	1.25	0.038
Phage T4 (TdR)	8.0	0.036
Poly dAT (^{3}H-T)	0.65	0.0013

Table III documents a more detailed experiment with radioactive pdAT.The data indicate that in spite of the relatively low uptake binding has occurred to the same sites that are involved with the binding and uptake of homologous DNA.This conclusion is based on the following observations:1)negative results were obtained when the radioactive pdAT was added to incompetent cells;2) there was no uptake of DNase-treated pdAT:3) homologous DNA added in excess simultaneously with the pdAT suppressed cellular uptake of the latter.

TABLE III

CHARACTERISTICS OF pdAT (^3H) BINDING

		Polynucleotide Bound in Nuclease-Resistant Form (CPM per ml)
Tube 1	1.0 ml Competent Cells + pdAT-^3H (5 x 10^7 viable units) (0.67 μg and 1.23 x 10^6 cpm)	2377
2	as above as above + Homologous DNA (70 μg)	230
3	as above pdAT-^3H, Pretreated with DNase (12 μg/ml, 30 mins, 30°C)	261
4	1.0 ml Incompetent Cells + pdAT-^3H	347
5	1.0 ml Competent Cells + dTTP-^3H (0.037 μg and 2.4 x 10^6 cpm)	122

4. DISCUSSION AND SUMMARY

Our results confirm and amplify the earlier findings of others concerning the requirement for double stranded secondary structure for cellular uptake of DNA (Lerman and Tolmach, 1957).The binding of molecules without free ends (i.e.,circular, double stranded) is surprising in view of the several lines of evidence suggesting linear entry of DNA molecules during genetic transformation. Possibly, these circular molecules undergo endonucleolytic scission at the cell surface at an early stage of the transformation process.Such activity has indeed been reported recently (Morrison and Guild, 1972). The activity of simple double stranded polynucleotides (pdA:dT, pdG:dC, etc.) in the binding assay indicate that the DNA binding and/or transporting system at the cell surface does not have an absolute requirement for specific base sequences in the donor molecules for its activity.However, substantial quantitative differences in the affinities of various polynucleotides may exist.The detection of these would require the preparation and comparison of homogeneous polynucleotides of high molecular weights since the affinity of competent cells is known to be the function of the molecular size (Cato and Guild, 1968). Experimental work along these lines is in progress in our laboratory.

ACKNOWLEDGEMENTS

These investigations were supported by grants from the US Atomic Energy Commission and the NATO Research Council.

5. REFERENCES

CATO, A. JR. & GUILD, W. R. (1968). J. Mol. Biol. 37:157.

LERMAN, L. S. & TOLMACH, L. J. (1957). Biochim. Biophys. Acta 26:68.

MORRISON, D. A. & GUILD, W. R. (1972). J. Bacteriol. 112:1157.

NAVA, G., GALIS, A. & BEISER, S. M. (1963). Nature, 197:903.

TOMASZ, A. (1969). Ann. Rev. Genetics, 3:217.

TOMASZ, A. & MOSSER, J. L. (1966). Proc. Natl. Acad. Sci. US, 55:58.

TOMASZ, A. & BEISER, S. M. (1965). J. Bacteriol. 90:1226.

TOMASZ, A. & ZANATI, E. (1971). J. Bacteriol. 105:1213.

TOMASZ, A., ZANATI, E. & ZIEGLER, R. (1971). Proc. Natl. Acad. Sci. US, 68:2627.

TOMASZ, A. (1966). J. Bacteriol. 91:1050.

TOMASZ, A. (1970). J. Bacteriol. 101:860.

ZIEGLER, R. & TOMASZ, A. (1970). Biochem. Biophys. Res.Commun., 41:1342.

THE MECHANISM OF BINDING OF THE COMPETENCE SUBSTANCE TO THE SURFACE OF NON-COMPETENT CELLS IN PNEUMOCOCCUS

MARGITA KOHOUTOVÁ

Department of Bacterial Genetics, Institute of Microbiology, Czechoslovak Academy of Sciences, Prague, Budějovická 1083, Czechoslovakia

ABSTRACT

Cells of genetically transformable strain R6bd of *Diplococcus pneumoniae* (streptomycin sensitive, sulfanilamide resistant) were grown for 50 minutes in the transformation medium and the cells were then concentrated by centrifugation and resuspension in a fresh medium to give 5×10^6 - 1×10^7 viable units per ml. Only under strictly regulated growth regiment is such a culture incompetent and can be induced to maximal competence by the added competence substance (CF). Experimental results have shown that not only the size but rather the physiological state of the inocula transferred into the transformation medium and their previous history decide whether a culture prepared as above will be completely incompetent or partly or largely competent.

Cells dividing for 5 - 6 generations in the pretransformation medium if transferred into the transformation medium at O.D. 0.12 - 0.18 (450 nm; Unicam spectrophometer SP 500) with a 20 fold dilution so as to give 1.5 - 5×10^6 viables per ml will be completely incompetent when grown and concentrated as above. The fundamental difference recorded with cultures transferred at various O. D. is in the concentration of the active CF needed for maximal induction of competence. The lower the O. D. of the transferred preculture the greater the concentration of CF required for induction (using identical count of exposed incompetent cells). With higher O.D. inocula also a greater induction yield of transformants (with identical concentrations of CF) can usually be achieved.

Cells dividing for only 2 - 3 generations in the pre-transformation medium (as happens with large and old inocula) give no correlation between the rise in O.D. and viables. On transfer of such a culture into the transformation medium partially or nearly fully competent cells are obtained even when using inocula with very low O. D. (corresponding to 1.3×10^6 viables per ml at zero time) and with identically treated culture as above. The biochemical and biophysical changes in the precompetent cell culture regulate the appearance of competence in the transformation medium.

Too high concentrations of partially purified CF give a low yield of transformants. The more physiologically mature the incompetent culture used the smaller the trigger required for the induction of competence by CF. The story seems to be very similar to the action of various concentrations of lysozyme in stimulation or lowering the yield of competent cells in B. subtilis. Recently we found that glycine at low concentrations can induce full competence in a low competent culture. The results seem to support the idea that the competent cell is a temporal spheroplast.

Phytohemagglutinins are plant proteins which interact specifically and reversibly with cell surface carbohydrates and glycoproteins. They cause animal cells to clump. Those substances bind to Sephadex dextran and are eluted from Sephadex in a similar way as is CF from Streptococcus and Pneumococcus. Supposing that both kinds of substances will also bind specifically and reversibly to similar or identical receptor sites on bacterial cell surfaces, the binding of CF would then be inhibited by the presence of agglutinins. In our work with various phytohemagglutinins we really found a great inhibition of induction of competence by CF in their presence. With high concentrations of phytohemagglutinins agglutination of non-competent pneumococcal cells was found.

Possible conformational surface changes introduced by binding of CF to surface receptor sites similar to those introduced by binding of phytohemagglutinins might lead to a series of reactions which would enable the cell to absorb the DNA molecules.

1. INTRODUCTION

For the adsorption of phages genetically determined specific receptors are required on the surface of the host cells. Conjugation is dependent on the presence of sex dependent fimbriae. In transformation and transfection naked donor DNA is attracted by the competent cells to which it adsorbes first reversibly followed by an irreversible binding. We know that some kind of mechanism exists in the competent cell which leads to a change of structural conformation of double-stranded donor DNA to a single-stranded form (Lacks, 1962), possibly at the time when DNA is bound to the cytoplasmic membrane in a DNase-insensitive form. Synapsis of single-stranded parts of donor DNA to a homologous part of the recipient genome (also possibly in a single-stranded replicative form),integration by recombination and phenotypic expression of the newly acquired genetic trait follow - as integral steps of the transformation reaction.

2. THE COMPETENT CELL CULTURE

The competent cell or better the competent culture which responses as a unit in Pneumococcus or Streptococcus when under certain conditions a 100 per cent synchronous competence develops in a few minutes, is a mature cell culture able of gene exchange (Ottolenghi & Hotchkiss, 1960), equipped with all required steps for the expression of the new trait.

Now what are the factors responsible for such a very abrupt development of competence in a cell culture?

McCarty, Taylor and Avery in 1946 at Cold Spring Harbour Symposium suggested "that it is not unreasonable to suppose that less extensive reversible alterations at specific sites of the surface of pneumococcal cells can result from enzymatic action, and that these alterations make possible the adsorption or penetration of the specific deoxyribonucleic acid".

I believe that this definition put forward at that early time of transformation experience and knowledge still holds and seems to express perfectly the competent state.

How far are we today to define competence in molecular terms?

René Thomas found competent cells of Pneumococcus to be more sensitive to osmotic shock (1957) than the non-competent cells and permeable for large molecules (Thomas, 1955). The competent cells seemed then to be changed in the cell wall structure. Towards a partial or fully spheroplastic state of "competent" (Iliashenko, Ditjatkin & Danillichenko, 1968) E.coli cells in transformation and transfection indicates the work of Kaiser and Hogness (1960), Chargaff, Schulman and Shapiro (1957), the penicillin or lysozyme prepared spheroplasts and recently glycine-spheroplasts (Taketo, 1972) used for infection with viral DNA. In B.subtilis lysozyme-spheroplasts prepared from competent cells are a good model for penetration of transforming (Tichý & Kohoutová, 1967, 1968; Hirokawa & Ikeda, 1966) and transfecting DNA (Tichý, Rytíř & Kohoutová, 1968).

In my early work I found that transformable pneumococcal strains isolated in my laboratory exhibited a very high nuclease activity leaking out from the cells (Kohoutová, 1965, 1967a,b) and they seemed to produce a much more active autolytic enzyme (unpublished results) than the highly transformable sub-lines of the R36A strain from Avery's laboratory (Avery,Mac Leod & McCarty,1944).The hypertonic media which were used by us then to obtain more regular results,improved 10 - 100 times the transformation frequency. This could be explained by stabilization of the unstable competent state of our recipient cells which might have been more spheroplastic than optimal. At the same time, the excess of endonuclease activity found in our strains was inhibited in the presence of hypertonic media (Kohoutová,1965,1967a,b).

Highly transformable strains of B. subtilis contain more N-acetyl galactosamine than the poorly transformable ones (Young, Spizizen & Crawford, 1963) and the galactosamine content of the cell wall increases with growth reaching a maximum at the time of maximal competence (Young, 1965). Young supposes that an increased substitution of galactosamine on terminal phosphorus group of teichoic acid could decrease the net negative charge of the cell wall and facilitate penetration of DNA (Young, 1967) . The modification of the cell wall by the autolytic enzyme N-acyl muramyl-L-alanine amidase with its maximal activity at maximal competence might produce focal relaxation and gaps in the cell wall and thus enable the penetration of DNA(Young, Tipper & Strominger, 1964).

From the very fine immunological work from the laboratory of Dr. Beiser (Nava, Galis & Beiser, 1963) it is known that a new antigen is connected with competence.

Jensen & Haas (1963) were first to show that competent cells of B.subtilis exhibit a reduced capacity to adsorb methylene blue. This was confirmed by Pakula, Ray & Spencer(1970) using competent Streptococci. The authors also found that competent Streptococci are highly precipitable at low pH which indicates a change of the surface charge.

3. COMPETENCE SUBSTANCE

Goodgal & Herriott (1961) found that on washing competent cells of H. influenzae competence is lost, but may be regained by readdition of washing fluids. A new approach was open to the study of competence when Pakula, Piechowska, Bańkowska & Walczak (1962) found that sterile supernatants of transformable Streptococcus strain contain a factor which induces competence in the non-competent Wicky cells. The finding of Pakula was soon proved to be correct. Tomasz and Hotchkiss (1964) isolated the competence factor from the surface of competent pneumococcal cells. In 1965 at the Mendel Symposium in Prague we described the substance in sterile filtrates of Pneumococcus where it accumulates after the peak of competence has passed(Kohoutová & Málek, 1966). Charpak & Dedonder (1965) found a competence factor both in B. subtilis supernatants and in extracts from the surface of cells. The work from the laboratory of Dr. Dobrzański on Streptococcus (Dobrzański & Osowiecki, 1966) and the work of Akrigg, Ayad & Barker (1967) on B. subtilis supported the importance of this substance in induction of competence. Osowiecki, Nałecz & Dobrzański (1969)found that the competence substance from Streptococci is positively charged. The same was found in our laboratory with competence substance from Pneumococcus (unpublished results). The competence substance on the competent cells may thus alter their negative charge and facilitate adsorption and penetration of DNA.

Most workers found that the competence substance is sensitive towards proteolytic enzymes. But there is no definite agreement as to the sensitivity of competence substance towards temperature,although the Polish results (Dobrzański & Osowiecki, 1966) and ours (Kohoutová & Málek,1966; Kohoutová, Braná & Holubová, 1968; Lipavská, Braná & Kohoutová, 1972) agreed very well. In both laboratories it was found that the substance is relatively temperature resistant. Highly purified competence substance from Streptococcus (Leonard & Cole, 1972) was also found to be thermoresistant in variance with the observation of Tomasz & Mosser (1966).Dr. Barnhart (1967) found a low molecular weight product of cell metabolism in sterile filtrates from H.influenzæ culture to be connected with competence stimulating activity. He proposed it to be an amino acid or a relatively low molecular weight polypeptide or an oligodeoxyribonucleotide. That competence substance is a small (Osowiecki, Nałecz & Dobrzański,1969; Leonard & Cole, 1972), dialyzable, basic compound (unpublished results from my and Dr. Dobrzański's laboratory; Leonard & Cole, 1972) has also been found.

According to Tomasz, the expression of competent character in a pneumococcal culture must be preceeded by the endogenous production of small quantities of the activator which initiates the self-catalytic process of activation. The author compares his activator substance to a hormon-like cell product which is responsible together with another extracellular factor, an inhibitor, for the regulation of the transient and synchronous expression of competence. The inhibitor which coincides with the termination of the competent phase of the culture inactivates the activator and prevents a further expression of competence(Tomasz, 1965, 1966). Such inhibitor has not been found by other workers (Pakula, 1965; Perry & Slade, 1966) and has never been found in my laboratory.

Now in my laboraotry we found that before the competence peak and at the peak we cannot find any competence substance in the supernatant and if so, then only a minimal amount at the peak of competence (Kohoutová, Holubová, Braná & Tichý, 1970). If a depletion of the activator would occur by the inhibitor found in the supernatant after the competence levels off, as Tomasz proposes, we would not find the highest level of competence substance 40 - 60 minutes after the peak of competence has passed. We also found in the isolation procedures of the competence substance from the surface of competent cells that the substance is at the peak of competence maximally bound to the surface of the cells (not yet published). And further, cells which have gone through the peak of competence in a synchronous culture are no more inducible by the added competence substance (Kohoutová & Málek, 1966; Kohoutová, Braná & Holubová, 1968; Kohoutová, Holubová, Braná & Tichý, 1970).

From these results it seems that competence is connected with the production of an endogenous competence substance which is genetically determined as shown by the results of Dobrzański, Nałecz & Rajchert-Trzpil (1970) with transformation of the marker of the competence factor, and is bound loosely or reversibly to the surface of cells at the competent state. The loss of competence in a cell culture is connected with the extrusion of the active substance from the competent cells into their surroundings.

4. PRECOMPETENT CELL CULTURE REGULATES THE APPEARANCE OF COMPETENCE

To prepare a physiologically incompetent culture for induction of competence from a genetically transformable strain is not easy. One can use the transformation medium (NS-medium, Kohoutová, 1965) devoid of bovine serum albumin but then very high amounts of competence substance have to be used for maximal induction of competence.One can also use a low pH transformation medium as suggested by Tomasz & Mosser(1966),but on resuspension of such grown cells into NS-medium of pH 7.8 very often spontaneous high competence develops. The best way how to prepare an incompetent culture is a strictly regulated growth regiment in the pretransformation medium (P-medium, Kohoutová, 1965).

Experimental results have shown that not only the size (Thomas, 1955; Taylor,1955; Kohoutová,1965) but rather the physiological state of the inocula transferred into the transformation medium and their previous history decide,whether a culture after 50 minutes growth in the NS-medium and after concentration so as to give $5 \times 10^6 - 1 \times 10^7$ viable units per ml will be completely incompetent or partly or largely competent.

In Fig. 1 we can follow three growth curves in the P-medium.The increase in optical density (O.D.)and in the viable count in the P-medium is strictly dependent on the size and the physiological state of the inoculum(brought in from the blood-P-medium).

In Fig.1a we can see a growth curve which started with 4×10^6 viable units per ml from a two hours and 30 minutes old culture from the blood-P-medium. This culture was able to go through 6.5 generation cycles.Such a culture if transferred into the NS-medium at O.D. 0.12 - 0.18 with a 20-fold dilution so as to give $1.5 - 5 \times 10^6$ viables per ml will be completely incompetent when grown and concentrated as above. The fundamental differences recorded with cultures transferred at various O.D. is

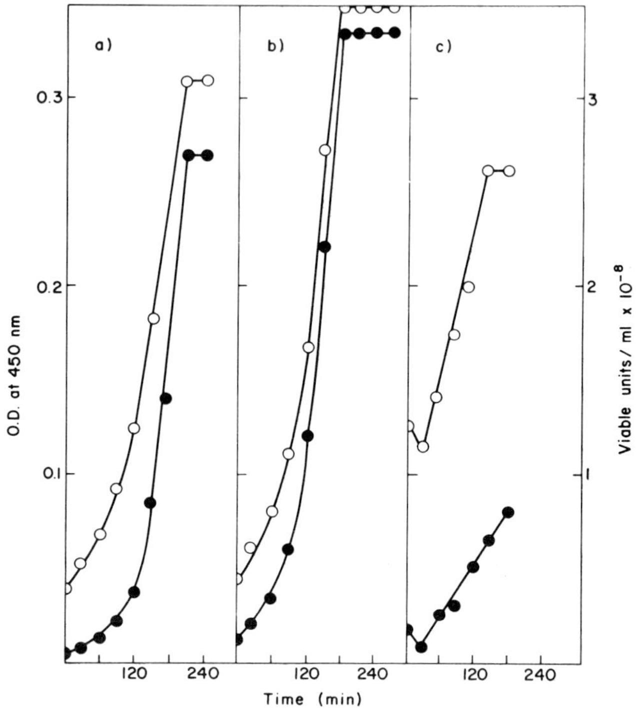

Fig. 1 - Optical density (O.D., Unicam Spectrophotometer SP 500) and growth (number of viable units per ml)of the R6bd strain of Pneumococcus in the pre-transformation medium (P-medium) in dependence on the inoculum culture from the blood-P-medium. •—• viable count; o—o O.D.

in the concentration of the active substance needed for maximal induction of competence. The lower the O.D. of the transferred preculture the greater the concentration of competence substance required for induction (using identical counts of exposed incompetent cells). With higher O.D. inocula also a greater induction yield of transformants (with identical concentration of active substance) can usually be achieved as will be shown later in Fig. 4.

In Fig. 1b the culture started with 1.2×10^7 viable units from a three hours old inoculum from the blood-P-medium and went through 4.5 generation cycles.

In Fig. 1c growth started with 1.8×10^7 viable units per ml from a more than four hours old inoculum from the blood-P-medium culture, and the culture was able to go only through 2.5 generation cycles. In such a culture there is no accordance between growth and O.D., indicating that a part of the culture is being represented by dead cells. One can also see a drop in O.D. and viables in the first 30 minutes of development of growth and, therefore, a number of cells transferred were not able to divide any more and represent also dead cells. In addition,one does not find here the logarithmic proper but only the linear increase of growth and of O.D..On transfer of such a culture into the transformation medium partially or nearly fully competent cells are obtained even when using inocula with very low O.D. (corresponding to 1.3×10^6 viables per ml at zero time)and with identically treated culture as above.

The biochemical and biophysical changes in the precompetent cell culture (and not only the cell size) regulate the ap pearance of competence in the transformation medium.

In Fig. 2 we see the time dependence of the appearance of the peak of competence in the NS-medium on the O.D.and on the size of the inoculum. The transformation medium was inoculated with a culture from the P-medium (Fig. 1b) with a 20-fold dilution.One has to keep in mind that with the increase in O.D. does not only increase the cell count and the peak of competence appears earlier but at the same time the physiological state of the culture as a whole changes.

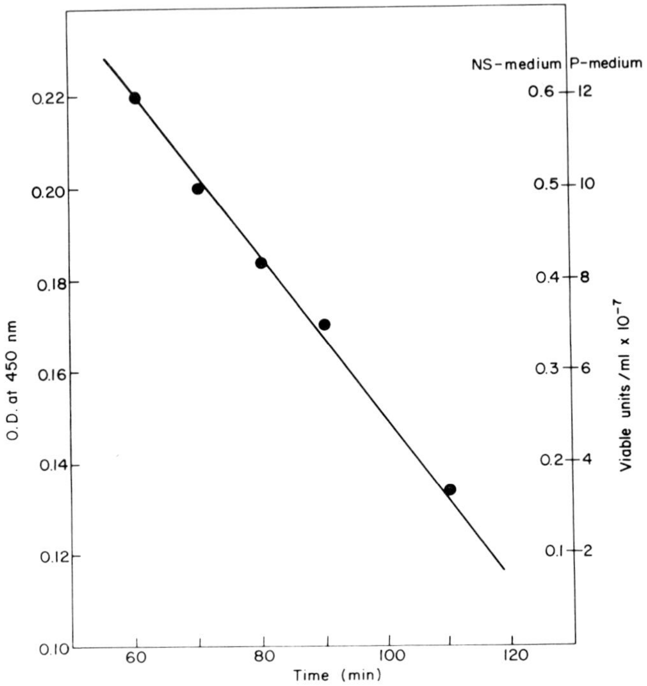

Fig. 2 - Time at which competence peak appears in the transformation medium. An inoculum culture from the P-medium as seen in Fig.1b)was transferred at various O.D. and 20 times diluted.

5. INDUCTION OF COMPETENCE IN INCOMPETENT CELLS GROWN FROM INOCULA OF VARIOUS PHYSIOLOGICAL STATE

Fig.3 shows the induction of competence in a 50 minutes old culture grown from a precompetent inoculum culture of O.D. 0.12 (Fig. 1a). Various cell concentrations are exposed to saturating concentrations of the active substance(0.2 ml of filtrate per system, 1.2 ml). When such a physiologically unmature preculture is used one can see that there exists an optimal number

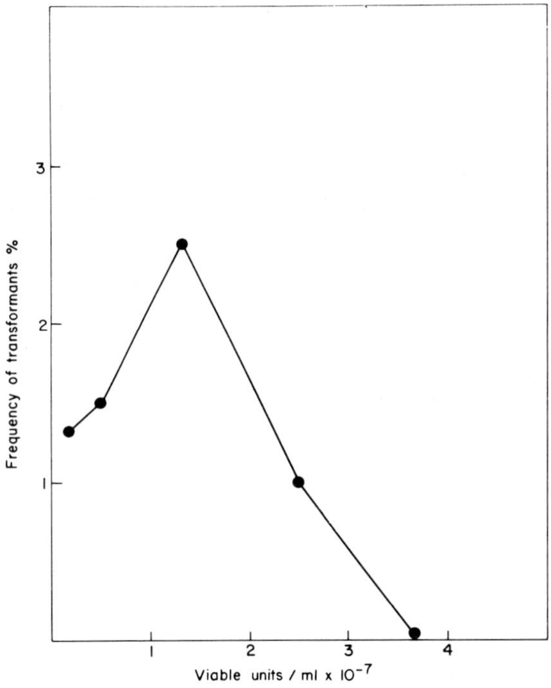

Fig. 3 - Induction of competence. Various concentrations of incompetent cells, grown for 50 minutes in the transformation medium, inoculated with a P-medium culture at O.D. 0.12 (as seen in Fig. 1a) were exposed to saturating concentrations of the competence substance.

of incompetent cells (1.2×10^7 per ml) in which maximal induction of competence (2.5%) can be achieved. With lower numbers of cells there seems to be a lack of receptor sites for the competence substance and, therefore, the frequency of transformants falls down to 1.3% with 3.3×10^6 recipients, while with 3.6×10^7 cells per ml of the same physiological state there seems to be a lack of the active substance to saturate the receptor sites. With a purified competence substance which we had available at that time and which was not at saturating concentration even for incompetent cells prepared from NS-medium cultures inoculated with more physiologically mature inocula we could not induce competence with any number of given cell concentrations.

Fig. 4 shows the induction of competence by sterile filtrate (a), by partially purified competence substance (b) and

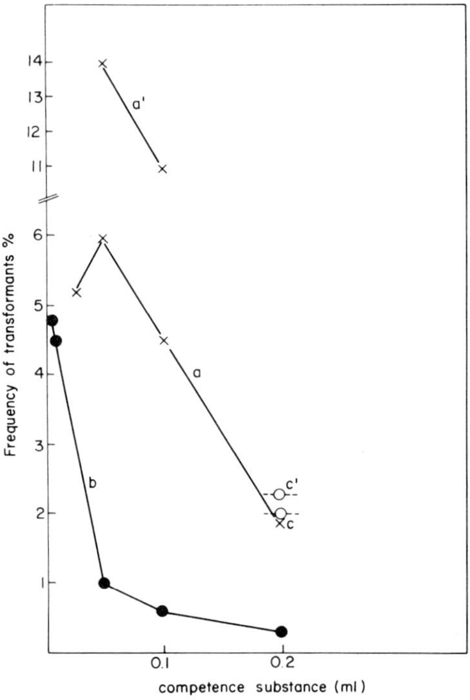

Fig. 4 - Induction of competence by sterile filtrate (a, a') ; by partially purified (b) and by highly purified (c, c') competence substance. In a, b, c an inoculum culture of O.D. 0.15 and in a', c' of O.D. 0.18 (Fig. 1a) was used. No of recipients: 3×10^7 and 1.5×10^7, respectively.

by highly purified substance (c) (Lipavská, Braná & Kohoutová, 1972) in a 50 minutes old culture when an inoculum of O.D. 0.15 (Fig. 1a) was used. The induction yield was here 6.0, 4.6 and 2.0% of transformants, respectively. If an inoculum of O.D. 0.18 (Fig. 1a) was transferred using the same filtrate (a') and the purified competence substance (c') 14% and 2.6% of transformants, respectively, were achieved. We can also see from the figure that a lower concentration of active substance is needed and a higher induction of competence can be achieved when a more physiologically mature pretransformation inoculum from the P-medium was used. The purified competence substance at non-saturating concentration (as mentioned above) could induce here 2.0 and 2.6% of transformants, respectively, while when a pretransformation inoculum culture with O.D. 0.12 (Fig. 3) was used no induction of competence could be achieved.

We can also see from Fig. 4 that high concentrations of the competence substance are inhibitory to the induction of competence. The more physiologically mature was the inoculum culture used for growth of the incompetent cell culture the smaller was the trigger required for the induction of competence by the competence substance. The story seems to be very similar to the action of various concentrations of lysozyme in stimulation or lowering the yield of competent cells in B. subtilis (Prozorov, 1965). Recently we found that glycine (approximately 0.03M at final concentration) can induce full competence in a low competent cell culture. The results seem to support the idea that the competent cell is a temporary spheroplast.

Could not then the binding of the competence substance to the cell receptor sites bring about also a temporary or reversible spheroplastic state? But what is the mechanism of binding of the active substance to the cell surface structure? Does it bind to the cytoplasmic membrane as Ziegler & Tomasz (1970) suppose or to some cell wall structure(s)? What are the receptor sites for the competence substance?

6. INHIBITION OF INDUCTION OF COMPETENCE BY THE COMPETENCE SUBSTANCE IN THE PRESENCE OF PHYTOHEMAGGLUTININS

Phytohemagglutinins are plant proteins which interact specifically and reversibly with cell surface carbohydrates and glycoproteins. They cause animal cells to clump. They bind to Sephadex and are eluted from it in a similar way as is the competence substance from Streptococcus and Pneumococcus (experiences from Dr. Dobrzański's laboratory,personal communication and confirmed in my laboratory).Supposing that both kinds of substances will also bind specifically and reversibly to similar or identical receptor sites on bacterial cell surfaces,the binding of the competence substance would then be inhibited by the presence of agglutinins. The determinants recognized by some of the agglutinins seem to be N-acetyl-D-glucosamine (Burger & Goldberg, 1967; Inbar & Sachs, 1969) in others N-acetyl-D-galactosamine (Lis, Sela, Sachs & Sharon, 1970) on the surface membranes of cells. Some evidence suggests that exposure of the agglutinin receptor sites is intimately connected with cell division and growth control (Burger, 1969; Fox, Sheppard & Burger, 1971).

While our experiments were in progress Doyle & Birdsell (1972) published the work about the interaction of agglutinin Concanavalin A with the cell wall of B. subtilis.

Table I shows the inhibition of induction of competence by the competence substance (sterile filtrate) in the presence of small amounts of various agglutinins using saturating concentration of the active substance. Agglutinin Phaseolus coccineus which does not bind to Sephadex exhibits a lower inhibition than the agglutinins Pisum sativum and Lathyrus sativus which do bind. At subsaturating concentrations of the active substance the lower the amount of the Phaseolus agglutinin the greater inhibition of induction of competence was recorded.

TABLE I

INHIBITION OF INDUCTION OF COMPETENCE BY VARIOUS PHYTOHEMAGGLUTININS (PA)

Competence substance Sterile filtrate(ml)	Induction: Frequency of transformants %	PA	Concentration µg/system	Inhibition of induction %
0.2*	10	Phaseolus Lathyrus	0.1 - 0.5	11 24
0.05* - 0.2*	10	Phaseolus Lathyrus	1	0 22.5
0.2*	4.3	Phaseolus Pisum Lathyrus	1 - 100	7 27 34
0.1**	1.3	Phaseolus	1 10 100	84 48 20

* Saturating concentrations of competence substance; 1.0 - 1.3 x 10^7 recipients per ml

** Subsaturating concentration of competence substance; 2.6 x 10^7 recipients per ml

Fig. 5 shows the inhibition of induction of competence in the presence of high concentrations of Phaseolus and Lathyrus agglutinins using a saturating concentration of the competence

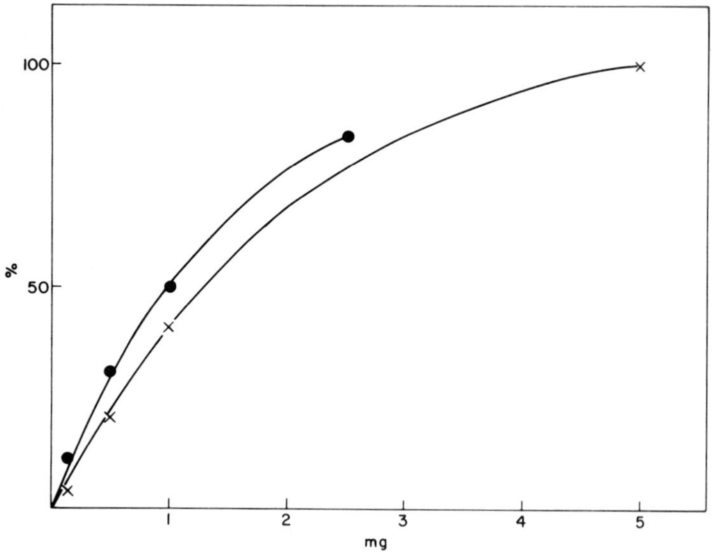

Fig. 5 - Inhibition of induction of competence in incompetent cells (4.5 x 10^6 per ml) in the presence of high concentrations of phytohemagglutinins using a saturating concentration of the competence substance. x——x : Phaseolus coccineus; •——•: Lathyrus sativus.

substance. At the highest concentrations 4 - 5 mg per system (1.3 ml) both phytohemagglutinins inhibit the induction of competence nearly 100% , then the inhibition falls down. In the presence of high concentrations of Lathyrus agglutinin the pneumococcal recipients are agglutinated and a heavy flocculation appears which also disappears at low concentrations (below 10 µg). With Phaseolus agglutinin no flocculation is apparent but a cloudy suspension appears at the highest concentrations and disappears from 500 µg downwards.

When subsaturating concentrations of the competence substance were used an unusual inhibition curve was observed. In the presence of high concentrations of agglutinin Pisum sativum (Fig. 6) an inhibition of induction of competence comparing to

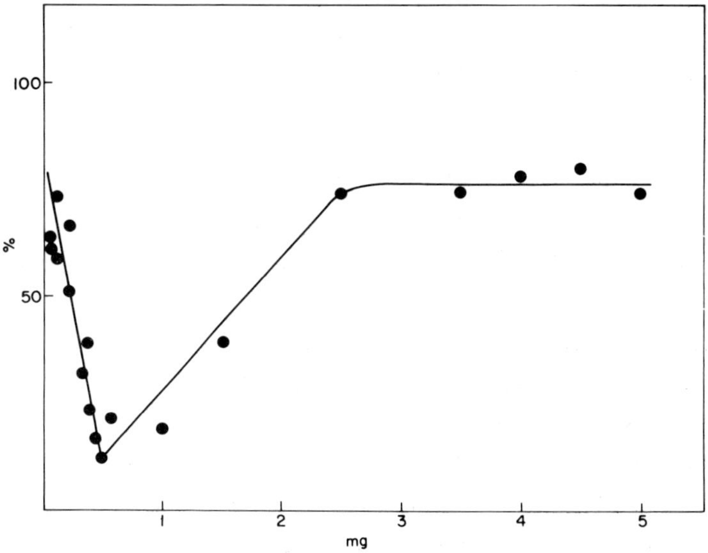

Fig. 6 - Inhibition of induction of competence in incompetent cells (1×10^7 per ml) in the presence of phytohemagglutinin Pisum sativum using a subsaturating concentration of the competence substance.

the above mentioned agglutinins was observed. Also strong flocculation of recipient cells appeared comparable to the Lathyrus agglutinin. But again with low concentrations, this time from 300 μg - 25 μg per system a second phase of inhibition was recorded. As the concentrations of the agglutinin decreased the inhibition of induction of competence increased. The elucidation of this second phase of inhibition has to wait for further experimental data.

7. DOES THE BINDING OF THE COMPETENCE SUBSTANCE TO THE CELL WALL RECEPTORS LEAD TO A TEMPORARY SPHEROPLASTIC STATE?

Now allow me to hypothetize on the ground of some of these still preliminary but I think interesting data which might bring us a step nearer to the elucidation of the mechanism of competence in transformation.

On the basis of possibly similar or identical receptor sites for some of the phytohemagglutinins and the pneumococcal competence substance on the glycopeptidic structure of the cell wall of Pneumococcus (or other transformable strains), and this glycopeptidic structure will be species and also strain specific, is it not plausible to suppose that the competence substance while binding to its specific receptor sites introduces some conformational changes which could lead to something like a fine agglutination? Thus introduced conformational changes might then trigger an autolytic enzyme activity (Young, 1967) which would produce local gaps for penetration of DNA already adsorbed to the positively charged competent cells. The DNA could thus pass into the interior of the cell bound by the endonuclease of the "unwindase" type described in our recent paper (Lipavská, Braná & Kohoutová, 1972;Kohoutová, Holubová, Braná e Tichý, 1970). DNA bound irreversibly on the cytoplasmic membrane to the endonuclease could then be stepwise unwound and synapse with the homologous part of the recipient genome.

Now on this ground we can better explain the results I have shown before. The physiologically younger inoculum culture used for growth and preparation of incompetent recipients used for induction of competence the greater amount of the competence substance required (a greater trigger has to be performed) for the conformational change which is connected with the appearance of competence - and even then maximal induction does not always occur. The more physiologically mature inoculum culture is used (some of the competence substance and more receptor sites have already been synthesized or exposed) the less competence substance is required for maximal induction.Too high concentrations of the competence substance might introduce such great conformational changes that the reversible state will change into the irreversible-one and the repair of the cell wall is no more possible.

A certain density of cells per ml is also necessary (Tomasz, 1966) not only because of an optimal concentration of the receptor sites (of glycopeptidic nature)needed for the binding of the competence substance but the contact of cells for a certain kind of agglutination process might be required.

I think that it is good to mention that Ephrussi-Taylor and Freed (1964) found an arrest in lysine incorporation into trichloracetic acid-insoluble residues coincident with the appearance of the competent state in Pneumococcus.The authors suggested that competence might be associated with an increase in an autolytic enzyme which specifically cleaves out lysine from the cell-wall complex and the synthesis of the peptide of the cell wall is momentarily inhibited.

The just proposed specific binding of the competence substance to the glycopeptidic structure of the cell wall with the introduced conformational change might be responsible for the limited autolysis and the temporary halt in the cell wall synthesis mentioned above.

ACKNOWLEDGEMENTS

The author is indebted to Dr. J. Kocourek from the Department of Biochemistry, Charles University for the generous gift of the phytohemagglutinins which were isolated and purified in his laboratory.

8. REFERENCES

AKRIGG, A., AYAD, S. R. & BARKER, G. R. (1967). Biochem.biophys. Res. Comm. 28:1062.

AVERY, O. T., MACLEOD, C. M. & MCCARTY, M. (1944). J.Exptl. Med. 79:137.

BARNHART, B. J. (1967). Biochem. biophys. Acta, 142:465.

BURGER, M. M. & GOLDBERG, A. R. (1967). Proc. Nat. Acad. Sci. Wash. 57:359.

BURGER, M. M. (1969). Proc. Nat. Acad. Sci. Wash. 62:994.

CHARGAFF, E., SCHULMAN, H. M. & SHAPIRO, H. S. (1957). Nature. 180:851.

CHARPAK, M. & DEDONDER, M. R. (1965). Compt. Rend. Acad. Sci. 260:5638.

DOBRZAŃSKI, W. T. & OSOWIECKI, H. (1966). Bull. Acad. Polon.Sci. 14:765.

DOBRZAŃSKI, W. T., NAŁECZ, J. & RAJCHERT-TRZPIL, M. (1970). Mol. Gen. Genetics. 108:312.

DOYLE, R. J. & BIRDSELL, D. C. (1972). J. Bact. 109:652.

EPHRUSSI-TAYLOR, H. & FREED, B. A. (1964). J. Bact. 87:1211.

FOX, T. O., SHEPPARD, J. R. & BURGER, M. M. (1971). Proc. Nat. Acad. Sci. Wash. 69:244.

GOODGAL, S. H. & HERIOTT, R. M. (1961). J. gen. Physiol.44:1201.

HIROKAWA, H. & IKEDA, Y. (1966). J. Bact. 92:455.

ILIASHENKO, B. N., DITJATKIN, S. J. & DANILEICHENKO, V. V. (1968). Biochim. biophys. Acta, 161:259.

INBAR, M. & SACHS, L. (1969). Proc. Nat. Acad. Sci. Wash. 63:1418.

JENSEN, R. A. & HAAS, F. L. (1963). J. Bact. 86:73.

KAISER, A. D. & HOGNESS, D. S. (1960). J. Mol. Biol. 2:392.

KOHOUTOVÁ, M. (1965). J. gen. Microbiol. 38:211.

KOHOUTOVÁ, M. & MÁLEK, I. (1966). Proc. G. Mendel Symp. Mutat. Process. The Physiology of Gene and Mutation Expression. ed. by M. Kohoutová & J. Hubáček, p.195.Praha: Academia.

KOHOUTOVÁ, M. (1967 a). Fol. Microbiol. 12:311.

KOHOUTOVÁ, M. (1967 b). Fol. Microbiol. 12:316.

KOHOUTOVÁ, M., BRANÁ, H. & HOLUBOVÁ, I. (1968). Biochem.biophys. Res. Commun. 30:124.

KOHOUTOVÁ, M., HOLUBOVÁ, I., BRANÁ, H. & TICHÝ, P. (1970). Fol. Microbiol. 15:183.

LACKS, S. (1962). J. Mol. Biol. 5:119.

LEONARD, C. G. & COLE, R. M. (1972). J. Bact. 110:273.

LIPAVSKÁ, H., BRANÁ, H. & KOHOUTOVÁ, M. (1972). Fol. Microbiol. 17:331.

LIS, H., SELA, B., SACHS, L. & SHARON, N. (1970). Biochim. biophys. Acta, 211:582.

MCCARTY, M., TAYLOR, H. E. & AVERY, O. T. (1946). Cold Spr.Harb. Symp. Quant. Biol. 11:177.

NAVA, G., GALIS, A. & BEISER, S. M. (1963). Nature, 197:903.

OSOWIECKI, H., NAŁECZ, J. & DOBRZAŃSKI, W. T. (1969). Molec. Gen. Genetics, 105:16.

OTTOLENGHI, E. & HOTCHKISS, R. D. (1960). Science, 132:1257.

PAKULA, R., PIECHOWSKA, M., BAŃKOWSKA, E. & WALCZAK, W. (1962). Acta Microbiol. Polon. 11:205.

PAKULA, R. (1965). J. Bact. 90:1320.

PAKULA, R., RAY, P. & SPENCER, L. R. (1970). Can. J. Microbiol. 16:345.

PERRY, D. & SLADE, H. D. (1966). J. Bact. 91:2216.

PROZOROV, A. A. (1965). Dokl. Akad. Nauk SSSR, 160:472.

TAKETO, A. (1972). J. Biochem. 71:507.

TAYLOR, H. E. (1955). Advan. Virus. Res. 3:275.

THOMAS, R. (1955). Biochim. biophys. Acta, 18:467.

THOMAS, R. (1957). Biochem. J. 66:38P.

TICHÝ, P. & KOHOUTOVÁ, M. (1967). Fol. Microbiol. 12:384.

TICHÝ, P. & KOHOUTOVÁ, M. (1968). Fol. Microbiol. 13:317.

TICHÝ, P., RYTIŘ, V. & KOHOUTOVÁ, M. (1968). Fol. Microbiol. 13:510.

TOMASZ, A. & HOTCHKISS, R. D. (1964). Proc. Nat. Acad. Sci. Wash. 51:480.

TOMASZ, A. & MOSSER, J. L. (1966). Proc. Nat. Acad. Sci. Wash. 55:58.

TOMASZ, A. (1965). Nature, 208:155.

TOMASZ, A. (1966). J. Bact. 91:1050.

YOUNG, F. E., SPIZIZEN, J. & CRAWFORD, I. (1963). J. Biol. Chem. 238:3119.

YOUNG, F. E., TIPPER, D. J. & STROMINGER, J. L. (1964). J. Biol. Chem. 239:3600.

YOUNG, F. E. (1965). Nature, 207:104.

YOUNG, F. E. (1967). Nature, 213:773.

ZIEGLER, R. & TOMASZ, A. (1970). Biochem. biophys. Res. Comm. 41:1342.

GEL ELECTROPHORESIS OF PROTEIN CELL EXTRACTS FROM STREPTOCOCCUS WICKY TREATED AND NOT TREATED WITH COMPETENCE FACTOR

PAWEŁ G. FUCHS
JOLANTA ZAJDEL
WŁADYSŁAW T. DOBRZAŃSKI

Department of Molecular Genetics,
University Medical School, Warsaw

ABSTRACT

Competence occurs in Wicky strain only after treatment of this strain with S. Challis competence factor (cpf). The role of the cpf in competence is unknown.

Ranhand et al. (J. Bacteriol. 104 : 360, 1970) suggested that cpf treatment of Wicky cells is followed by the synthesis of a new protein which is detectable by gel electrophoresis in cell phenol extracts. Ranhand et al. used in their experiments a crude preparation of the cpf.

It was decided to perform similar experiments, but using highly purified cpf preparation and additional control systems.

In the conditions used in these experiments, the appearance of a new protein band in cell extracts could not be demonstrated. Some changes in gel patterns occuring after cpf treatment of Wicky cells are discussed.

Development of competence and transformability in group H Streptococci depends on synthesis by recipient or contact of recipient with a competence provoking factor (cpf). Group H Streptococcal strain Wicky does not produce cpf,but it is possible to induce the competence in this strain by treatment with cpf preparation originating from another group H streptococcal strain Challis (Pakula & Walczak, 1963). The biological function of cpf in DNA uptake remains unknown, as yet.

Recently, Ranhand et al.(1970)reported,that cpf treatment of Wicky cells is followed by synthesis of a new protein, which is detectable in the acidified-phenol extracts of the cells by mean of gel electrophoresis. In view of these important data it was decided to perform similar experiments, extended by some new elements.

Ranhand et al.described the electrophoretic pattern of the extract from erythromycin-resistant Wicky cells, pretreated with not purified cpf preparation and incubated in Brain Heart Infusion for about 60 min. Controls were performed with extract of Wicky cells, not-treated with cpf and treated simultaneously with cpf and chloramphenicol (CAP),which is known to inhibit the development of competence. An additional peak, though very small, was observed only in the cpf-treated pattern. This pattern was characterized by a rather high number of protein bands, as was also the case with the patterns of control samples.

In the present work wild Wicky strain was used and incubation was performed in HE medium (Piechowska & Shugar, 1967). To eliminate the possible effect of impurities,cpf rather highly purified by CM-cellulose column chromatography was used (Osowiecki, Nałecz & Dobrzański, 1969).After cpf treatment, the cultures were incubated for 30 min. only and not for 60 min., since this period seems to be too long and new events can occur during that period.After 30 min.,the cultures were immediately strongly cooled. Similarly to Ranhand et al., cpf not-treated, cpf-treated and cpf+CAP treated cell extracts were tested. Moreover, it was also checked, what changes the patterns of cell extracts may undergo, when treated with cpf in vitro and what changes in the patterns will follow cpf-cells contact for 5 min.only - a period as a rule too short for synthesis of a new protein.

Acidified phenol extracts were prepared by the procedures identical with those reported by Ranhand et al., in some cases only extraction was longer by 15 min. The electrophoresis was performed in acidic poliacrylamide gel system, composed of stacking gel, containing 3% acrylamide in formic buffer, pH 4.5 and the separating gel, containing 15% acrylamide in formic buffer,pH 3.5.The buffer,used for preparation of the gels contained 8M urea. The gels were photopolymerized in 0.5 x 7.0 cm tubes in the presence of 0.0005% riboflavine and tetramethylethylenediamine(TEMED, 20 µl per 10 ml of the gel).Samples of the extracts (250-400 µg protein) were layered on the top of the gels and electrophoresis was run at a constant current of 4 mA per tube, using a pyronine dye as a marker. Gels were stained with Amido Black 10B, and tracings of protein bands were obtained with Carl Zeiss (Jena) Densitometer.

Since a certain fluctuation of the results could be anticipated, all determinations were performed parallely, each extract being tested three times.

Fig. 1 shows the result of in vitro treatment with cpf of an extract from cpf not-treated cells, in comparison with the patterns of not-treated cells and cpf itself. The pattern of purified cpf preparation consists of some peaks on the left side only.They are grouped together and compact within the region. In

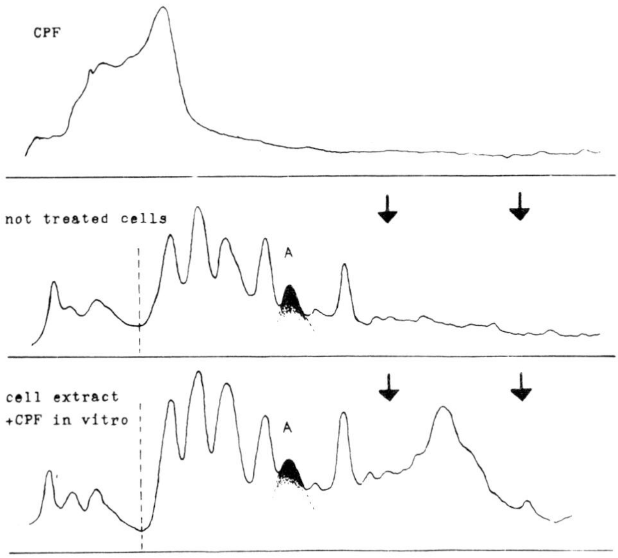

Figure 1

the extract of not-treated cells (middle pattern), peak denoted A is always rather low, both absolutely and relatively, as compared with neighbouring peaks and no well visible peaks can be found on the right side, in the region marked with the arrows. The bottom pattern refers to the extract of cells, treated with cpf *in vitro*. The peak A is also low. However, a certain part of cpf proteins seems to be localized on the right side, thus reversely to the electrophoretic patterns of cpf alone.

Fig. 2 presents a comparison of electrophoretic patterns, obtained from Wicky cells, treated in a different way. Top pattern (a) shows the extract of cells treated with cpf for 30 min. Control, not treated with cpf is illustrated by the pattern

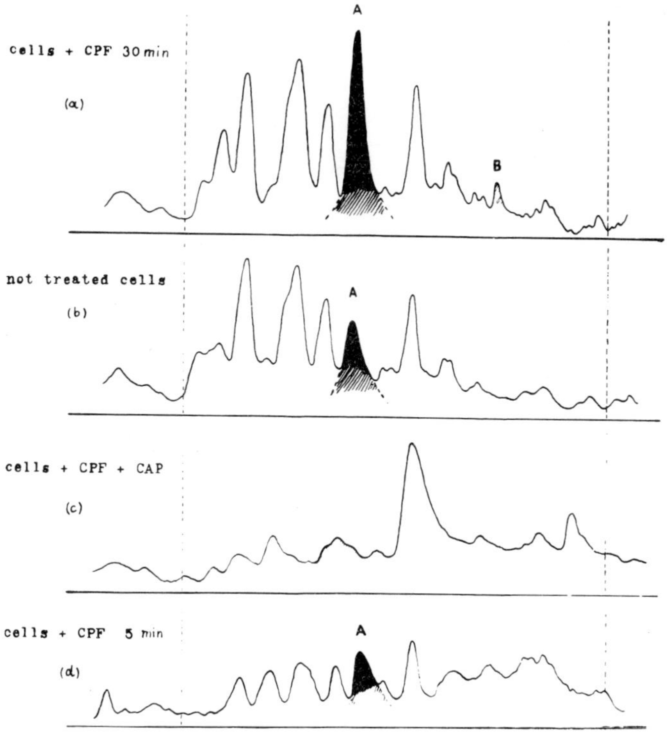

Figure 2

(b). Compared with control, the pattern (a) shows a considerable increase in peak A, both absolutely and in relation to the neighbouring peaks, what might be a result of stimulation of this protein(s) synthesis by cpf components. Some changes and in particular the appearance of marked increase of peak denoted B can also be observed within a region, where some cpf components were localized after in vitro reaction (see Fig. 1). The pattern (c) refers to cells treated with cpf for 5 min. only. A certain increase in peak A can be seen, though lower than in the top pattern, as well as changes on the right side, where cpf components are grouped. This pattern seems to be intermediate in character between control (b) and sample (a) treated with cpf for 30 min.

Pattern (d) presents a control, in which the cells were treated simultaneously with cpf and CAP. Peak A is low and perhaps slight changes can be seen on the right side, but not in the peak B region. As Ranhand et al. we also found that in this control much less protein was extracted, thus the sample contained smaller amount of cell protein. This fact decreases the validity of the control, because the sensivity of the method may be too low to demonstrate small protein fractions in the extract. The additional peak, found by Ranhand et al. seems to represent such a small protein fraction.

What are the conclusions from these experiments?

1. In cpf-treated cells the appearance of a new protein was not demonstrated. If peak B even represents a new protein, it cannot be considered as cpf-induced, since some cpf components in controls are localized in the same region.

2. A new observation is, that in the reaction *in vitro*, and possibly also *in vivo*, considerable part of cpf components combine only with defined cell proteins. We were not successful, however, in determining where specific activity of cpf is localized.

3. Absolute and relative increase in peak A might be the reflection of a selective stimulation of the synthesis of this pre-existing protein(s) by some components of cpf preparation, for example, by derepression. However, such interpretation may be perhaps challenged by the result, obtained with cells, treated with cpf for only 5 min. In this sample a certain small increase in peak A also occured, although in fact there was no time for protein synthesis - unless it starts very rapidly.

4. Differences between Ranhand et al. and our results may be due to various reasons, e.g. some differences in experi-

mental details and/or by the use in these experiments of the highly purified cpf preparation.

5. The above discussed observations brought some new data,which seems to indicate,however, that such a kind of experiments cannot yet bring results that could be safely interpreted. For example, the experiments should be performed with absolutely pure cpf and we should be able to identify its activity in the gels.

REFERENCES

OSOWIECKI, H., NAŁECZ, J. & DOBRZAŃSKI, W. T. (1969). Molec. gen. Genet. 105:16.

PAKULA, R. & WALCZAK, W. (1963). J. gen. Microbiol. 31:125.

PIECHOWSKA, M. & SHUGAR, D. (1967). Acta biochim. pol. 14:349.

RANHAND, J. M., THEODORE, S. T. & COLE, M. (1970). J. Bact. 104:360.

SOME OBSERVATIONS ON THE MECHANISM OF COMPETENCE WITH THE USE OF BLACKBURN STRAIN GROUP H STREPTOCOCCUS AND ITS MUTANTS

PIOTR CEGŁOWSKI
BOHDAN J. STAROŚCIAK
WŁADYSŁAW T. DOBRZAŃSKI

Department of Molecular Genetics,
University Medical School,Warsaw

Similarly to Challis and Wicky strains, <u>S. blackburn</u> belongs to serological group H Streptococci.

In our previous communications we reported that:

1. competence and transformability in Challis and Wicky strains depend most likely on the presence and/or proper localization in the recipient of the factor which inactivates <u>in vitro</u> transforming DNA/iF/, and not only on the activity of the competence factor /cpf/. The iF activity was found in alkaline and acid cell extracts. The available data indicate that crude preparations of iF contain one or more nucleases.

2. non-transformable Blackburn does produce cpf, and its cell extracts do not contain iF.

In the present work,transformable mutants were isolated from the wild-type Blackburn strain /cpf^+,iF^-/ after mutagenesis with nitrosoguanidine. In cellular extracts of these mutants iF was found, which is not present in extracts of wild-type cells. Thus, the transformable mutants of the Blackburn strain are cpf^+, iF^+ . This finding confirms the view on the role played by iF in competence.

From the transformable Blackburn mutants cpf^+ iF^+, a cpf^- iF^+ mutant was isolated after mutagenesis. In this mutant competence and transformability may be induced by cpf treatment. The non-transformable cpf^- iF^- mutant was isolated in a similar way from the wild strain cpf^+ iF^-.

The UV sensivity of all the isolated mutants is similar to that of the wild-type Blackburn strain and the transformable Challis strain.

Using the above mentioned Blackburn strains which repre sented all the possible combinations of cpf and iF activity, the following observations on the mechanism of competence were performed:

1. the part played by cpf and iF in irreversible and reversible binding of isotope labelled DNA was investigated,

2. the role of cpf and iF in spontaneous release of transforming DNA by cells was studied,

3. taking into consideration reports (Tomasz, A. and Zanati, E. 1971 J. Bact. 105:1213, Pakula et al. 1970, Can. J. Microbiol. 16:345) that in Diplococcus pneumoniae and in Challis and Wicky strains competent cells agglutinate at acid pH values, and incompetent cells do not, the effect of cpf and iF on the ability of Blackburn strains to undergo acid agglutination and also the role of cpf in the agglutination of Challis cpf^+ and cpf^- strains were tested,

4. Ravin A. W. and Ma, M. (1972, J. Bact. 109:616), reported that heating a competent Challis strain for a short time at 48°C considerably lowers its ability to bind DNA irreversibly. Taking into consideration this report, the influence on cpf and iF of 20-minute heating at 48°C in the competent Blackburn strain was tested,

5. the effect of chloramphenicol on the activity of cpf and iF in the transformable Blackburn strain was examined.

THE FIMBRIATED STATE AND COMPETENCE IN GENETIC TRANSFORMATION OF THREE *MORAXELLA* SPECIES

L. O. FRØHOLM

K.BØVRE

National Institute of Public Health,Oslo, Norway
and
Kaptein W. Wilhelmsen og Frues Bakteriologiske Institutt, University of Oslo, Rikshospitalet, Oslo, Norway

ABSTRACT

Moraxella nonliquefaciens, *M. bovis* and *M. kingii* appear with different colony types showing low frequencies of intertype variation. One type has spreading/corroding (SC) colonies with, most often numerous, fimbriae demonstrable by electron microscopy ; the other type has non-spreading/non-corroding (N) colonies with no or only few and late appearing fimbriae. Cells of the latter type regularly reveal incompetence or low to medium competence in streptomycin resistance transformation, whereas SC-type cells show high to medium competence. Extensive investigations on pairs of variants in a large number of strains have always shown the SC representative to have a T/E ratio (number of transformed cells/exposed colony-forming units) of about 100 to more than 10^6 fold higher than the corresponding N variant. SC variants with lower T/E ratio than some N variants of other strains are known,however. Fimbriae seem to give cells of moraxellae a definite advantage for genetic transformation, but the mere presence of a few fimbriae in the population is not sufficient to ensure high competence, and the cells can apparently be transformed at a low T/E ratio without having detectable fimbriae.Centrifugation in linear density gradients of Urografin (Renografin) often gives several bands of each cell line and occasionally the variants of one strain band at different densities. The more competent variant does not,as a rule, band at a lower density than the less competent one.

1. INTRODUCTION

In the species M.nonliquefaciens, M.bovis and M.kingii we have detected that genetically high-competent bacteria are distinguished by their colony morphology and constant possession of fimbriae (s. pili). These findings will be briefly described in the following with references to published work, and preliminary data on density gradient centrifugation of the variants will be presented. We have reasons to believe that a similar correlation between competence and fimbriation will be found in other transformation systems when looked for specifically. We will also point out how the fimbriae could possibly facilitate some steps in the transformation process by discussing recent work of others.

2. RESULTS

(a) Colony variation

Strains of the three species investigated have been shown to grow different types of colonies on human blood agar plates when cultivated in a very humid atmosphere (Bøvre & Frøholm, 1972*a*; Frøholm & Bøvre, 1972*a*; Henriksen, 1969; Henriksen & Bøvre, 1969). Plate I shows photographs of variants of M.bovis 4 as they appear after growth for two days at 33°. Part of the agar surface has been removed with a platinum loop to allow inspection of the agar beneath the colonies. One bacterial variant grows evenly circular colonies with a smooth top surface. No trace is usually left on the agar substrate after scraping. This is the N type, demonstrated in Plate I(a). The other colony type is usually quite different, as shown in Plate I(b). These colonies often cover a larger area than the N type, due to surface spreading, the edges

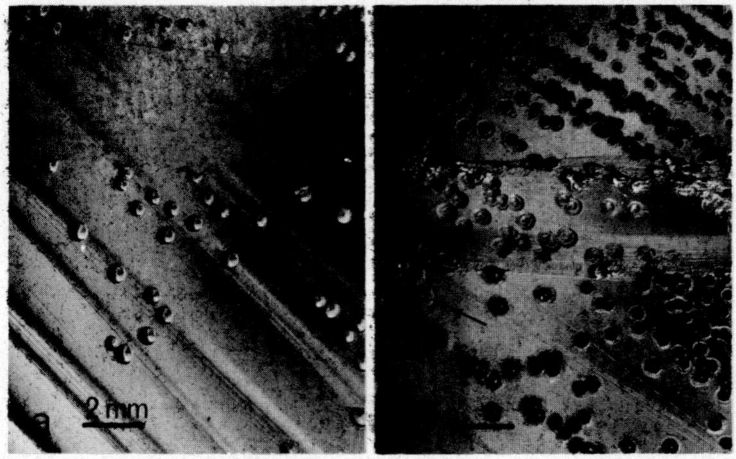

PLATE I - MACROPHOTOGRAPHY OF M. BOVIS 4 COLONIES ON AGAR

Bacteria were spread on human blood agar plates (Bøvre & Frøholm,1972a) and incubated in a humid atmosphere for two days at 33°. A Nikon F camera with PB-4 bellows focusing attachment combined with a telephoto lens (E. Leitz, Elmar f = 9cm 1:4) was used. Illumination was with a photolamp placed close to the side of the lens and the colonies were photographed at a primary magnification of 2:1 in areas where the light was reflected from the agar surface. Magnification 5:1.
(a) *M. bovis* 4 N-b with hemispherical, circular colonies, no spreading zone, and no corrosion left after scraping.
(b) *M. bovis* 4 SC-a. Colonies with wide zone of spreading, in some places under the agar level except for the peripheral edge (arrow), in other places partly above the agar layer (double arrow). The raised, central part of the colonies smaller than 4 N-b colonies grown strictly in parallel. Corrosion clearly visible corresponding to the central colony (deep pit) + spreading zone.

are often uneven with tonguelike projections and the surface is rough-looking. After scraping, a depression is left in the agar corresponding to the site of the colony. This is named the spreading and corroding type (SC). N is the abbreviation for the non-spreading and non-corroding type. The relative size of SC colonies (as compared to N)may vary with the species, strain,the cultural conditions and the number of passages after isolation from the parasitic stage,as may also the size of the spreading zones.Agar corrosion is a constant characteristic of the SC colony type in the three species of *Moraxella*, however.

In most of the strains thoroughly investigated the two morphological types are found to show variation into the other at a very low frequency. When more than 10,000 well separated colonies are studied,it is usually impossible to detect a single one that has changed morphology, indicating a frequency of spontaneous variants well below 10^{-4}. The analysis is complicated by presumed intra-colonial selection mechanisms(Bøvre & Frøholm , 1972*a*; Frøholm & Bøvre, 1972*a*).

(b) Electron microscopy of colony variants

When the bacteria from the agar surface are prepared for electron microscopy by negative staining, the situation is often found to be as shown in Plate II (see the legend for technical details): Cells of both types regularly grow as diplobacilli with some tendency to chain formation. The SC-type cells have filamentous appendages - to a varying degree - as a constant feature. The filaments are usually straight, of 60 - 70 A diameter, and of different length,from short pieces to several μ. In Plate II(a), *M. bovis* 4 SC-a shows a typical electron microscopical picture of SC cell lines. There seem to be several fimbriae coming from each cell in this line, often with a subpolar origin. A slight tendency to bundle formation is evident (arrow). There is some strain variation in the degree of fimbriation: they are not always seen easily near all cells in the preparations (Bøvre,

Bergan & Frøholm,1970; Bøvre & Frøholm, 1972a). The N-type cells usually show no fimbriae whatsoever - as was clearly the case with the *M. bovis* 4 N-b seen in Plate II(b) - but there may in other cases be a few in the preparations.These few fimbriae tend to increase in number with prolonged incubation and, typically, first appear later than the 5 - 7hr of growth which is usual for harvesting cells as recipients in our transformation experiments (Bøvre & Frøholm, 1972b). The phenomenon may possibly be due to intracolonial selection or regulation mechanisms.

The cells of the N type are slightly smaller than those of the SC type in the two variants of *M.bovis* 4 here shown. This has not been noticed in the other strains and variants studied earlier.

Both types of cells have a convoluted cellular surface and to a varying degree tubule-like projections of the surface layers are seen (double arrows).These cell wall extensions (Frøholm & Bøvre, 1972b) are not markedly different in the two types of cells.In Plate II(c) and II(d) the fimbriae and the cell wall extensions are shown at a higher magnification. In addition one can find granular material close to the cells by this method(not shown here). Accumulation of such material is sometimes a very striking feature of the N-type cells which then look as if they are embedded in a pile of fine grains making it difficult to define the cell surface accurately and also,of course, difficult to exclude entirely the presence of a few fimbriae - especially very short ones - close to the cell surface.

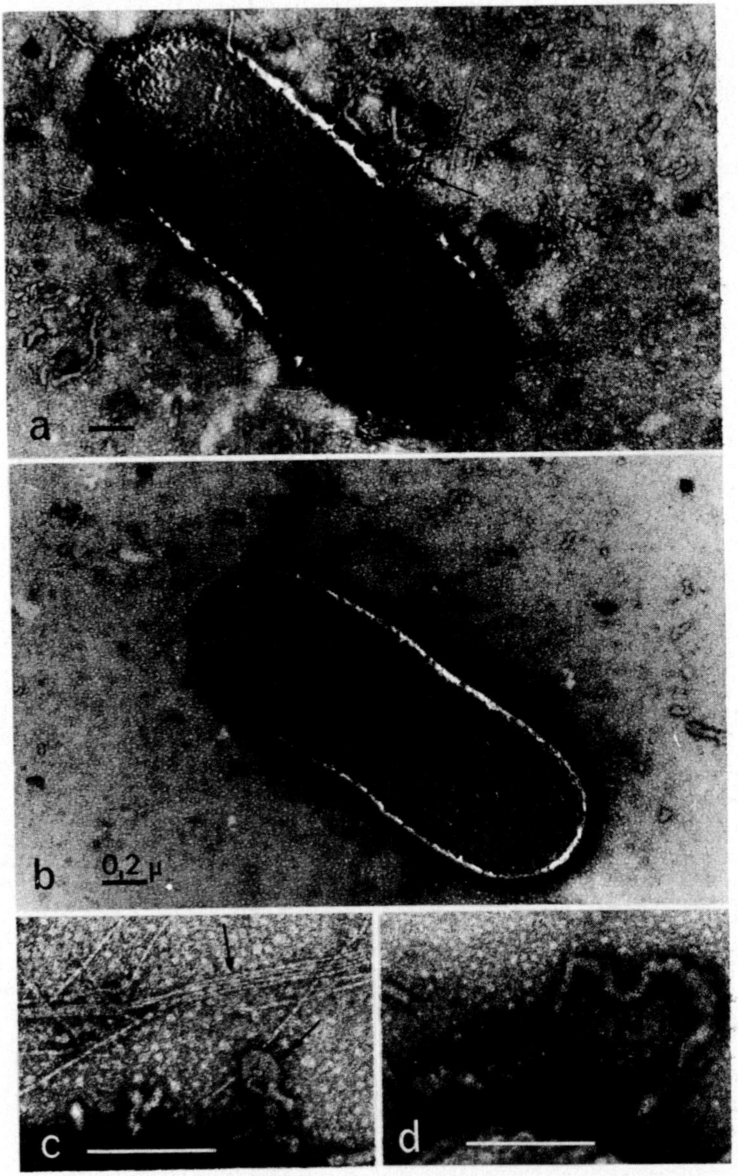

PLATE II - ELECTRON MICROSCOPY OF M.BOVIS 4 GROWN ON AGAR

Bacteria grown for 20 hr (see legend to Plate I for conditions) were harvested by scraping part of the agar surface

(c) Transformability of variants

When pairs of colony variants (= fimbriation variants) of the same strain were tested in streptomycin transformation experiments with the same batch of DNA (usually isolated from N-type cells of the same species)it was regularly found that the SC-type cells yielded from about 100 to over one million fold more transformants than the N-type recipient (Bøvre & Frøholm, 1972*b*). Up to almost five per cent of the exposed colony forming units of the SC type, in some strains, were transformed under these conditions of short time DNA exposure in solution terminated with DNase.Transformants from N-type recipients were often - but far from always - found to be enriched as spreading and

lightly with a straight end, sterile spatula. The material collected was spread by squeezing on a clean glass slide and shortly afterwards suspended evenly in negative stain solution(0.8 per cent (w/v) sodium silicotungstate,pH 6.5) by pipetting; the volume was adjusted to give a light greyish suspension. A small drop of this was transferred to a carbon coated,formvar covered copper grid and surplus liquid removed after about half a minute by touching the edge of the grid with a filter paper (Bøvre & Frøholm, 1972*a*). Magnification: (a) & (b) 35,000:1, (c) & (d) 100,000:1.
(a) *M. bovis* 4 SC-a with fimbriae (arrow), both solitary and in small bundles of two and three running in parallel. Cell wall extensions indicated with double arrow.
(b) *M.bovis* 4 N-b surrounded by globular and more extended structures (double arrow) presumably originating as cell wall extensions.
(c) Fimbriae and cell wall extensions of *M.bovis* 4 SC-a at higher magnification. The edge of the cell is seen in the lower part of the picture. The white line represents 0.2 μ.
(d) The edge of a cell of *M. bovis* 4 N-b showing cell wall extensions.
An optical differentiation method (Markham, Frey & Hills, 1963)was employed in (a) and (b)to visualize both the cell surface and the extracellular structures in the same print.

corroding colonies - possibly due to the presence of undetected variants in the recipient population. Table I shows the data obtained with the variants we have been able to isolate from the type strains of the three species investigated. No entirely non-fimbriated N cell lines were found among these strains, however. Still the T/E ratio is as low as 10^{-5} to 10^{-8} for the N variants while the SC ones are all above 10^{-2}.

Occasionally a typical SC cell line in other strains was found to have a relatively low competence - i.e. a T/E ratio of less than 10^{-4}. This is close to or even below the competence of certain N-type cell lines from other strains. However, the N-type cell lines isolated from the same strains were invariably found to be even less competent(by the factor range given above) or totally incompetent (Bøvre & Frøholm, 1972b).

(d) Density gradient centrifugation of moraxellae

We have done a few experiments to test the density of some of our variants in linear gradients of Urografin in order to compare with the well documented system of light,physiologically competent cells of Bacillus subtilis 168 (Cahn & Fox, 1968; Hadden & Nester, 1968; Singh & Pitale, 1967).

We have not encountered a similar situation with our bacteria, where we have pairs of stable variants differing considerably in level of transformability. No clear and consistent pattern can be found. Several of the cell lines give rise to two or more bands in the gradient as can be seen in Plate III. The SC variants may be lighter (not shown), of equal density (NCTC 7784) or heavier than the corresponding N variant (4177/66). M. kingii cells seem to be heavier than those of M. nonliquefaciens.

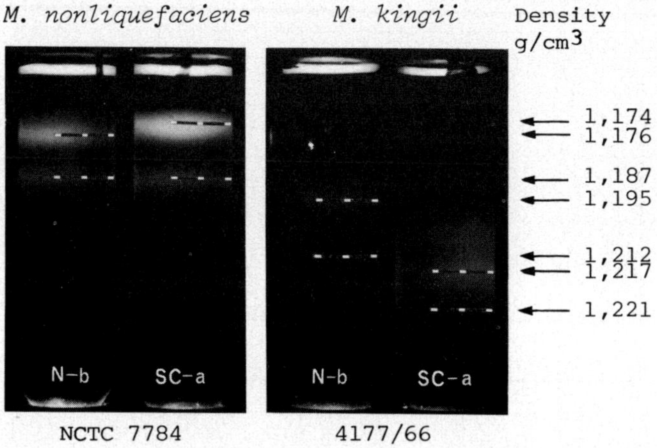

PLATE III - DENSITY GRADIENT CENTRIFUGATION OF MORAXELLAE IN UROGRAFIN

See Bøvre & Frøholm (1972a , 1972b) and Frøholm & Bøvre (1972a) for further descriptions of the strains and variants employed. Cells grown for about 24 hr were harvested directly in buffer (0.13 M NaCl, 0.01 M Tris-HCl, pH 7.5) by gently rubbing the agar surface with the straight part of a bent glass rod, followed by centrifugation for 10 min at 3,000 x g in an angle head rotor. The sediment was washed once by suspension in Urografin of density 1.10 g/cm^3 and centrifuged as above. The pellet was resuspended carefully in the same Urografin solution and 0.5ml put on top of a gradient obtained by layering 0.5 ml portions of the following Urografin solutions (density values given) through a canula introduced to the bottom of a 5ml centrifuge tube: 1.16; 1.17; 1.18; 1.19; 1.20; 1.21; 1.22; 1.23; 1.24. The gradient was allowed to stand 2 hr in the cold before adding the sample and immediately starting the run in a Spinco L50 centrifuge, rotor SW50.1, speed 30,000 rpm for 120 min, with cooling and without brake. Under these conditions the gradient proved to be linear over an extended part of the tube (checked by refractometry of fractions collected) and the bands reached apparent equilibria. Photography was done in a dark box with a 7 mm diameter bundle of light passing vertically down the tubes. All the specimens shown display a distribution of cells in more than one band. The densities of the bands indicated are given to the right as determined by measurement of refractive index (Cahn & Fox, 1968).

TABLE I

FIMBRIATION AND TRANSFORMATION COMPETENCE IN VARIANTS OF THE MORAXELLA NONLIQUEFACIENS, M. BOVIS AND M. KINGII TYPE STRAINS[a]

Species strain	Variant[b]	Degree of fimbriation[c]	Colony-forming units exposed to DNA per ml (E)	Transformed cells per ml (T)	T/E ratio
M. nonliquefaciens	N-b	(-)*	1.0×10^{9}	$<1.0 \times 10^{1}$	$<1.0 \times 10^{-8}$
4663/62 = ATCC 19975[d]	N-a	(-)	1.4×10^{8}	5.5×10^{3}	3.9×10^{-5}
	SC-a	++	1.2×10^{7}	3.0×10^{5}	2.5×10^{-2}
M. bovis	N-a	(-)*	2.0×10^{8}	4.5×10^{3}	2.3×10^{-5}
ATCC 10900	SC-b	++	1.7×10^{7}	3.8×10^{5}	2.2×10^{-2}
M. kingii	N-a	(-)*	3.4×10^{8}	$\leq 1.0 \times 10^{1}$	$\leq 2.9 \times 10^{-8}$
4177/66 = ATCC 23330	SC-a	++	3.0×10^{7}	3.4×10^{5}	1.1×10^{-2}

(a) Electron microscopy was performed as described in legend to Plate II and by Bøvre & Frøholm (1972 a). Quantitative streptomycin resistance transformation was employed as described by Bøvre (1964), with few modifications (Bøvre & Frøholm, 1972b). Transforming DNA originated from a resistant mutant of N-type cells of the recipient strain. For transformation experiments with other strains and variants of the three species, see text and ref. Bøvre & Frøholm, 1972b.

(b) N: non-spreading, non-corroding colony type. SC: spreading, corroding colony type. Labels a - b indicate cell lines (Bøvre & Frøholm, 1972a, Frøholm & Bøvre, 1972a).

(c) (-)*: fimbriae occasionally observed, but not in very young cultures (about 7 hr), as those used as recipients in transformation. (-): fimbriae occasionally observed, also in young cultures (about 7 hr). ++: most cells with several or numerous fimbriae.

(d) ATCC: American Type Culture Collection, Rockville, Maryland, U.S.A.

3. DISCUSSION

On the basis of the present experiments we believe it is correct to conclude that the presence of fimbriae by itself is not sufficient for the cells to be highly transformable, and a certain low level of transformability is often found with no fimbriae or at least very few per cell. But it seems beyond doubt that the presence of fimbriae in most instances gives the cells of the Moraxella species studied a clear advantage in terms of transformability. We have not yet had the opportunity to test for competence in the strict sense of only irreversible DNA binding. Possibly the specific role of fimbriae would prove even more clear in such tests. We have previously expressed (Bøvre & Frøholm, 1971) that we believe fimbriae will also be shown to be correlated with transformability in Neisseria gonorrhoeae.

In collaboration with Dr. Jyssum we have initiated cultural and electron microscopical studies on some of his competent and incompetent variants of N. meningitidis (Jyssum & Lie, 1965; Jyssum, 1971). These studies are not yet completed but it seems clear that the genetically competent variants usually are fimbriated to a much higher degree than the incompetent ones. In this system there is no difference between the colonies of the variants. It has, however, been detected that the fimbriated variants are able to form a pellicle when grown statically in broth (L. O. Frøholm, K. Jyssum & K. Bøvre, unpublished observations). Pellicle growth has previously been described as a characteristic of fimbriated bacteria (Bøvre, Bergan & Frøholm, 1970; Bøvre & Frøholm, 1972a; Duguid, 1968; Old & Duguid, 1970).

Whether the fimbriae themselves represent part of the uptake mechanism for DNA or another functional unit in the trans formation process, or merely reflect some underlying principle (operating without participation of fimbriae) cannot be decided by us on the basis of the present experiments. Also we cannot explain exactly how the fimbriae should function if they were part of an uptake mechanism.

The sex fimbriae (F pili) of *Escherichia coli* are responsible for cell contact during conjugation. They are by Brinton (1965) postulated to act as organs of transport for the bacterial chromosome from the donor (carrying the F fimbriae) to the recipient ("F-pili conduction model"). The same fimbriae do also represent sites of attachment for two different types of so-called male specific bacteriophages: the filamentous DNA phage (*e.g.* f1) attaches to the free ends of F fimbriae (Caro & Schnös, 1966) and the icosahedral RNA phage (*e.g.* MS2) attaches to the side. It is also postulated by Brinton (1965) that these phage nucleic acids are taken up through the fimbriae which must then be able to conduct DNA in both directions as well as RNA into the bacterium.

The "F-pili retraction model" has been proposed as an alternative explanation for the functioning of fimbriae (Curtiss, 1969; Marvin & Hohn, 1969). Quite recently Bradley (1972) has presented evidence in favour of this mechanism being in operation in *Pseudomonas aeruginosa* during infection with the bacteriophage PP7. Bradley has shown that the fimbriae in this bacterium are shortened upon attachment of the bacteriophage which finally is found located close to the cell surface. In connection with this proposed retraction, it may be worth mentioning that Mayer (1971) has found that the fimbriae of the starforming *Pseudomonas echinoides* contracts during star formation when the cells are drawn together by their polar fimbriae. A working model for this contraction has been published by Mayer & Schmitt (1971).

Our system has not yet allowed such experiments. We have, however, detected that our fimbriated bacteria show the peculiar phenomenon of "twitching motility" (Henrichsen, Frøholm & Bøvre, 1972). This kind of surface translocation explains the spreading zone found around colonies of fimbriated bacteria. The mechanism by which fimbriae give bacteria the ability to "twitch" is unexplained as yet,but some kind of contraction or retraction of fimbriae after attachment of their ends would offer one possible explanation. Similarly, contraction or retraction upon attachment of DNA could facilitate DNA uptake.

4. REFERENCES

BRADLEY, D. E. (1972). Biochem. Biophys. Res. Comm. 47:142.

BRINTON, C. C., Jr. (1965). Trans. N. Y. Acad. Sci. Ser. II. 27:1003.

BØVRE, K. (1964). Acta Path. Microbiol. Scand. 61:457.

BØVRE, K., BERGAN, T. & FRØHOLM, L. O. (1970). Acta Path. Microbiol. Scand. Sect. B. 78:765.

BØVRE, K. & FRØHOLM, L. O. (1971). Nature New Biol. 234:151.

BØVRE, K. & FRØHOLM, L. O. (1972a). Acta Path. Microbiol. Scand. Sect. B. 80:629.

BØVRE, K. & FRØHOLM, L. O. (1972b). Acta Path. Microbiol. Scand. Sect. B. 80:649.

CAHN, F. H. & FOX, M. S. (1968). J. Bact. 95:867.

CARO, L. G. & SCHNÖS, M. (1966). Proc. Nat. Acad. Sci. Wash. 56:126.

CURTISS, R. III. (1969). Annu. Rev. Microbiol. 23:69.

DUGUID, J. P. (1968). Arch. Immunol. Ther. Exp. 16:173.

FRØHOLM, L. O. & BØVRE, K. (1972a). Acta Path. Microbiol. Scand. Sect. B. 80:641.

FRØHOLM, L. O. & BØVRE, K. (1972b). J. Ultrastruct. Res. In press.

HADDEN, C. & NESTER, E. W. (1968). J. Bact. 95:876.

HENRICHSEN, J., FRØHOLM, L. O. & BØVRE, K. (1972). Acta Path. Microbiol. Scand. Sect. B. 80:445.

HENRIKSEN, S. D. (1969). Acta Path. Microbiol. Scand. 75:85.

HENRIKSEN, S. D. & BØVRE, K. (1969). Acta Path. Microbiol. Scand. 76:459.

JYSSUM, K. (1971). Acta Path. Microbiol. Scand. Sect. B. 79:265.

JYSSUM, K. & LIE, S. (1965). Acta Path. Microbiol. Scand. 63:306.

MARKHAM, R., FREY, S. & HILLS, G. J. (1963). Virology, 20:88.

MARVIN, D. A. & HOHN, B. (1969). Bacteriol. Rev. 33:172.

MAYER, F. (1971). Arch. Microbiol. 76:166.

MAYER, F. & SCHMITT, R. (1971). Arch. Microbiol. 79:311.

OLD, D. C. & DUGUID, J. P. (1970). J. Bact. 103:447.

SINGH, R. N. & PITALE, M. P. (1967). Nature, Lond. 213:1262.

FATE OF TRANSFORMING DNA

THE BIOLOGICAL SPECIFICITY OF HAEMOPHILUS ENDODEOXYRIBONUCLEASES WHICH ATTACK HETEROLOGOUS DNA

SOL H. GOODGAL
ROSA GROMKOVA

Department of Microbiology, School of Medicine
University of Pennsylvania, Philadelphia, Pennsylvania 19104

ABSTRACT

Specific endodeoxyribonucleases (endo-R) isolated from Haemophilus species have been purified and shown to attack and inactivate transforming DNAs of heterologous strains. The relative loss of transforming ability of endo-R treated DNAs depends not only upon the source of the enzyme but also on the recipient used in the transformation assay. Heterospecific transforming activity of some markers is higher when the recipient in transformation is the same strain as the source of the enzyme suggesting that endo R's play a specific role in excluding heterospecific DNA. Endo R's break DNA into specific segments some of which retain biological activity. Different endo-R affect different sites and give different patterns of survival for transforming markers in contrast to sonication or pancreatic DNAse treatment which have little or no specificity. The homology between different strains of Haemophilus is reflected in the survival of similar markers from DNAs treated with a specific endo-R. High molecular weight DNA used in heterospecific transformation between H. parainfluenzae and H. influenzae results in killing of the recipient cell. Treatment with endo-R reduces the size of the transforming DNA leading to a marked increase in the survival of the recipient and an increase in heterospecific transformation of certain markers. These data suggest that small segments of DNA can contain the sites for efficient recombination.

1. INTRODUCTION

The discovery by Smith's group of an enzyme of *Haemophilus influenzae* (endo-R) which breaks down foreign DNA but not its own raised some interesting possibilities for studying the phenomenon of restriction of DNA. Since the phenomenon affected a system in which transformation was well established one could approach the study of restriction from a number of points of view, including studies of phage restriction and heterospecific transformation.

In this paper we wish to consider the effects of endo-R enzymes from related Haemophilus species on the biological activity of heterologous DNAs and the ability of these DNAs after treatment to transform homospecifically and heterospecifically.

The primary effect of endo-R on heterologous DNA is to produce double strand breaks reducing the molecular weight of the DNA to an average size of approximately 10^6 daltons (Smith and Wilcox,1970).The segments produced by endo-R have been shown to have specific ends corresponding to a break in the middle of the nucleotide sequence $\frac{\text{G T Py Pu A C}}{\text{C A Pu Py T G}}$ (Kelly and Smith, 1970).

The existence of specific ends has been confirmed by treating DNA of *H. influenzae* with endo-R from *Haemophilus parainfluenzae* and sedimenting the segments produced in a sucrose gradient. The transforming activity of many markers cannot be detected after endo-R treatment, but those which retain some biological activity are found to sediment with a characteristic size (Goodgal and Gromkova,PNAS, in press). In general the amount of biological activity remaining after endo-R treatment is proportional to the size of the segment of DNA on which it is found. What is perhaps of considerable importance is the fact that a number of different endodeoxyribonucleases have been discovered in related Haemophilus species which attack DNA at different sites. Thus, it is now possible to generate segments of DNA of defined size which are amenable to studies of nucleotide sequence.

The ability of DNA to transform heterospecifically has been correlated with the relatedness of the strain of bacteria employed, and more to the point, with the genetic sequence homologies between their DNAs (Schaeffer, 1958). If endo-R is responsible for specific restriction of heterologous DNA then DNA treated in vitro with endo-R may reflect the same kinds of breaks as those expected in vivo. One could test this hypothesis by observing whether or not a relationship exists between heterospecific transformations and the residual activity remaining in a DNA after treatment with an endo-R. The test requires that the recipient utilized for transformation be the same as the source of the specific endo-R. With a number of strains available for study with their corresponding endo-R enzymes one can test a large number of systems for the role of endo-R in the restriction process.

(a) The efficiency of heterospecific transformation

In order to measure the effect of endo-R on heterospecific transformation it is necessary to determine the efficiency of heterospecific transformation. Table I presents the summary data from a number of experiments using H. influenzae, H. parainfluenzae, Haemophilus aegyptius, and Haemophilus influenzae Reid as recipients and donors of a number of markers tested homospecifically and heterospecifically. In all cases the percentage

TABLE I

EFFICIENCY OF DIFFERENT MARKERS IN HOMOSPECIFIC AND HETEROSPECIFIC TRANSFORMATION

Donor DNA	Marker**	Recipient			
		H. inf.	H. para	H. aeg.	H. inf. Reid
H.inf.*	Str	100	0.60***	35	80
	Nal	100	0.44	30	>100
	Dal	100	5.00	66	>100
	Kan	100	2.70	--	---
	Nov	100	0.08	1.7	>100
	Ery	100	10.0	30	>100
H.para*	Str	0.05	100	0.13	3
	Nal	0.3	100	--	--
	Dal	0.18	100	--	4
	Nov	0.001	100	0	0
	Ery	10	100	7.5	12
H.inf. Reid*	Str	40	0.25	25	100
	Nal	45	0.1	--	100
	Dal	60	1.0	90	100
	Nov	42	0.1	20	100
	Ery	45	10	40	100
H.aeg.*	Str	50	1	100	--
	Nal	70	0.1	100	80
	Dal	90	9	100	--
	Nov	60	0.1	100	--
	Ery	80	10	100	--

* Abbreviations: H. inf., H. influenzae; H. para, H. parainfluenzae; H. inf. Reid, H. influenzae Reid; H. aeg., H. aegyptius.

** Antibiotic resistance markers.

*** % efficiency relative to homologous transformation.

figures refers to the efficiency relative to the same marker transformed homospecifically, e.g. str-r DNA of H. influenzae used to transform H. parainfluenzae gave an efficiency of 0.6% relative to a str-r marked DNA of H. parainfluenzae. The DNA's were all used at concentration of 1 γ/ml or greater which is at least 10 x the concentration necessary for saturating the competent cells.

Leidy,Jaffee and Alexander (1962) from their studies on heterospecific transformation with H.influenzae, H.aegyptius, and H.influenzae Reid, concluded that these strains were closely related and that H. parainfluenzae was less related. Our data confirms the fact that markers transform with high efficiency between H. influenzae, H. influenzae Reid, and H. aegyptius and that toward these strains H. parainfluenzae is considerably poorer as both donor and recipient of transforming DNA.

(b) In vitro activity of endo-R enzymes on the transforming efficiency of Haemophilus DNAs in homospecific transformation

Marker efficiencies in Haemophilus transformation in general do not vary by more than a factor of two. However, after endo-R treatment there is a considerable variation in marker efficiencies. In order to correlate the effects of endo-R in vitro and in vivo we have examined the effects of endo-R on homospecific transformation on a number of markers. Table II summarizes the data obtained using the DNAs and endo-Rs from four different strains.

These data confirm our original finding (Gromkova and Goodgal, 1972) that endo-Rs can attack unrelated DNA but do not destroy the transforming activity of their own DNA. H. influenzae Reid was tested as a crude extract and found to have relatively little endo-R activity, however its own DNA was quite sensitive to the endo-R preparations from H. influenzae, H. aegyptius, and H. parainfluenzae. This was somewhat surprising since H. influenzae Reid and H. influenzae can act as good re-

TABLE II

IN VITRO ACTIVITY OF endo-R ENZYMES ON THE TRANSFORMING EFFICIENCY OF HAEMOPHILUS DNAs IN HOMOSPECIFIC TRANSFORMATION

DNA	Marker	endo-R enzyme*			
		H. inf.	H. para.	H. aeg.	H. inf. Reid
H.inf.	Str	100**	<1	<1	78
	Nal	100	200	<1	71
	Dal	100	<1	40	--
	Nov	100	90	<1	70
	Ery	100	<1	<1	--***
	Kan	100	<1	35	--
	Vio	100	<1	60	--
	Spec	100	<1	50	--
H.para.	Str	<1	100	<1	50
	Nal	10	100	<1	--
	Dal	<1	100	60	--
	Nov	10	100	<1	50
	Ery	<1	100	<1	50
H.inf. Reid	Str	<1	<1	<1	100
	Nal	30	22	<1	100
	Dal	10	<1	10	100
	Nov	50	17	2	100
	Ery	25	<1	1.5	100
H.aeg.	Str	<1	<1	100	---
	Dal	<1	7	100	---
	Nov	1	4	100	---
	Ery	<1	<1	100	---

* Procedure for endo-R assay: DNA was diluted to 25 μg per ml in Tris - Mg - mercaptoethanol buffer (6.6 mM each Tris buffer, pH 7.4, $MgCl_2$ and mercaptoethanol). To 0.5 ml of this material was added 5μ liters of purified enzyme and the mixture was incubated at 30°C. Samples were removed after 60 min. and examined for transforming activity. H.inf. Reid was tested as a crude extract (See Smith and Wilcox, 1970 and Gromkova and Goodgal, 1971).

** % activity remaining at 60 min.

*** Not tested.

cipients for each others transforming DNA. The same relationship was also found for H. aegyptius. The endo-R from H. aegyptius was inactive against H. aegyptius DNA but attacked the markers of H. influenzae and H. influenzae Reid, strains with which it readily exchanges markers by transformation. A general order of relatedness could be deduced from the level of survival of markers, e.g. H. influenzae Reid DNA was less susceptible to H. influenzae endo-R than H. aegyptius endo-R, and H. influenzae DNA was less susceptible to H. aegyptius endo-R than to H. parainfluenzae endo-R. Moreover,the patterns of inactivation were different for the three purified endo-R enzymes suggesting that these enzymes act at different sites in the DNA.

A considerable amount of evidence is now available to show that two endo-R preparations from different Haemophilus species can reduce the size of susceptible DNA more than either one alone and that the segments produced by one endo-R may retain biological activity for some markers but in subsequent exposure to a second endo-R this activity is destroyed. See Table III.

TABLE III

EFFECT OF SEQUENTIAL ENZYME ACTION ON H. PARAINFLUENZAE DNA

DNA	Marker	Enzyme	Recipient	
			H. para.	H. inf.
H. para	Ery-r	H. inf.-60 min	<1*	100
		H. inf. 60 min then H. aeg. 60 min	<0.1	<0.1
H. para	Ery-r	H. aeg. 60 min	<1	4
		H. aeg. 60 min then H. inf. 60 min	<0.1	<0.1

* % activity remaining

(c) Some markers have a preferential survival

One feature of the action of endo-R on a series of markers in different strains of Haemophilus is the fact that some markers are less readily attacked by specific endo-Rs, e.g. endo-R from H. aegyptius appears to spare the dal-r marker in the three DNAs tested while the endo-R from H. parainfluenzae is less active on the nal-r and nov-r markers than the other markers presented in Table I. These data may simply indicate basic genetic homology between these genetic characteristics, but also suggest a similarity in endo-Rs and their mechanism of action. An observation of some interest is the fact that the high level str-r markers are very sensitive to all endo-Rs and support the notion that the str-r loci represent alterations which affects a rather large segment of DNA.

In contrast to the effect of endo-R, transforming DNA which has been treated by shearing or sonication loses biological activity as a function of reduction in size, and for most markers there is an equivalent loss of activity for any give size DNA.

(d) The survival of markers heterospecifically in different recipients after endo-R treatment

The data presented in Table IV demonstrates clearly that the segments of DNA produced by endo-R in vitro are in most cases less efficient in heterospecific transformation than untreated DNA. (The data are presented in terms of the % survival of treated/untreated DNAs). However, if one examines the survival of markers in recipients from which the endo-R enzyme was obtained (compared to other heterospecific recipients) one finds

TABLE IV

EFFECT OF endo-R TREATMENT ON HETEROSPECIFIC TRANSFORMATION

DNA	Enzyme	Recipient	Marker				
			Str	Nov	Dal	Ery	Nal
H.inf.	H. para	H. inf.	<1*	90	<1	<1	200
		H. para	<1	100	---	13	---
		H. aeg.	---	---	---	---	---
		H.inf.Reid	<1	100	---	<1	---
H.inf.	H. aeg.	H. inf.	<1	<1	40	<1	<1
		H. para	<1	---	50	1	---
		H. aeg.	<1	---	63	<1	---
		H.inf.Reid	---	---	---	---	---
H.para	H. inf.	H. para	<1	10	<1	<1	10
		H. inf.	<1	---	---	100	---
		H. aeg.	<1	---	---	1	---
		H.inf.Reid	1	<1	5	10	---
H.para	H.aeg.	H. para	<1	<1	60	<1	<1
		H. inf.	<1	---	---	10	---
		H. aeg.	<1	---	---	90	---
		H.inf.Reid	---	---	---	---	---
H.inf. Reid	H. inf.	H. inf.	<1	50	10	17	---
		H. para	<1	100	1	<1	---
		H. aeg.	<1	100	25	70	---
		H.inf.Reid	<1	50	10	25	---
H.inf. Reid	H.para	H. inf.	<1	22	<1	<1	---
		H. para	<1	100	<1	<1	---
		H. aeg.	<1	70	<1	<1	---
		H.inf.Reid	<1	17	<1	<1	---
H.inf. Reid	H. aeg.	H. inf.	<1	<1	25	<1	---
		H. para	<1	25	15	1	---
		H. aeg.	<1	80	25	1	---
		H.inf.Reid	<1	2	10	1.5	---
H.aeg.	H. inf.	H. inf.	12	50	10	15	---
		H. para	<1	60	3	5	---
		H. aeg.	<1	20	<1	<1	---
		H.inf.Reid	1	10	1.5	2	---
H.aeg.	H.para	H. inf.	<1	50	<1	1	---
		H. para	<1	100	2.5	<1	---
		H. aeg.	<1	4.8	7	<1	---
		H.inf.Reid	<1	2	<1	<1	---

* % activity remainning at 60 min.

that in general the survival is better in the recipient supplying the endo-R. For example, the endo-R of *H. influenzae* attacking *H. aegyptius* DNA shows 12% survival of the str-r marker in *H. influenzae*, but less than 1% survival in *H. aegyptius* or *H. influenzae Reid* recipients. This was also true for the dal-r and ery-r markers. These data suggest that the endo-R double strand breaks produced in vitro are similar to those made in vivo. Furthermore, it is clear that severe restrictions are imposed on the transforming DNA as a result of its decrease in size.

The marker specificity of endo-Rs discussed above is also apparent in Table IV. The str-r marker is very sensitive to all endo-Rs except homologous enzyme. On the other hand the dal-r marker is resistant only to the *H. aegyptius* endo-R but is sensitive to all the others.

Keeping in mind the fact that endo-R from *H. influenzae* attacks *H. aegyptius* and *H. influenzae Reid* DNA's even though these DNA's are very efficient in transformation of *H. influenzae* (Table I), one can conclude that these DNA's are not modified through covalent bonds which protect them against attack by endo-R, and that the *H. influenzae* cell must have the ability to prevent the action of its own endo-R so that *H. aegyptius* and *H. influenzae Reid* markers are not inactivated during transformation. It is unlikely that *H. aegyptius* or *H. influenzae Reid* DNA are markedly reduced in size during transformation of *H. influenzae* since the complex or large str-r markers transform efficiently. The str-r marker encompasses a DNA length of approximately 3-4 x 10^6 daltons (Bendler, Ph.D. Thesis, 1968), and is very sensitive to endo-R in vitro. Randolph and Setlow (1972) have shown that following sonication there is a good correlation between reduction in transforming activity and lack of integration of DNA.

At this point one can conclude that endo-R of H. influenzae reduces the size of H. aegyptius or H. influenzae Reid DNA in vitro and consequently limits their transforming activity because of this size reduction but that in vivo the ability of endo-R to attack DNA is in some way controlled. Three major possibilities for this control are 1) that endo-R is itself altered;2) there is a protective material which prevents endo-R from attacking the DNA; and 3) the structure of DNA after uptake is changed so that it cannot be attacked.

(e) The ery-r marker from H. parainfluenzae transform efficiently heterospecifically as a relatively small segment of DNA.

In previous studies we have shown that endo-R from H. parainfluenzae can attack Haemophilus influenzae DNA and reduce its average size to 10^6 daltons. Nevertheless, some segments are preserved which contain biological activity and in general this activity can be found in segments whose size is greater than 10^6 daltons. Indeed, the larger the segment of DNA in which a marker is found the greater the amount of activity retained. One might therefore anticipate that the biological activity found after treatment of H. parainfluenzae DNA by H. influenzae endo-R would be in the relatively large segments of DNA.

From the information presented below we are led to conclude that this is not always the case; and that endo-R may actively serve to preserve markers for transformation by reducing the size of DNA.

In Table IV one can observe that all the ery-r marker activity of H. parainfluenzae is retained after treatment with

H. influenzae endo-R if the activity is measured in H. influenzae, and from Table II it is seen that the heterospecific transforming efficiency of H. parainfluenzae ery-r marker in H. influenzae is 10% (Nickel, 1964; Beattie and Setlow, 1970), i.e. it has a relatively high efficiency compared to other markers.

In transformations using DNA from H.parainfluenzae and H. influenzae cells there is an inactivation of H. influenzae cells due to exposure to H. parainfluenzae DNA (Steinhart and Herriot, 1968; Beattie and Setlow, 1970). We have confirmed these observations by examining the viability of H. influenzae competent cells exposed to DNA fractions from a sucrose gradient of H. parainfluenzae DNA. We have found that large segments of DNA produced few transformants heterospecifically and the viability of the cells was reduced whereas the same DNA homospecifically yielded high levels of transformants. Since the viability of the cell population was reduced by only 25% and the surviving population contained very few transformants for markers like str-r one can conclude that it is the potential heterospecific transformants which were killed. This argument is reinforced when one considers that the smaller segments of H.parainfluenzae do not kill the recipient H. influenzae cells and that the residual transforming activity is located in the sucrose gradient in a position corresponding to smaller segments of DNA. Furthermore, one observes that the ery-r marker transforms heterospecifically with high efficiency with small segments of DNA, but poorly with small segments homospecifically. This observation becomes all the more interesting when one examines the effect of endo-R on the transforming efficiency of the ery-r marker. The effect of H. influenzae endo-R on H. parainfluenzae DNA is to reduce its size to an average of 10^6 daltons, and the residual transforming activity of the str-r, ery-r, and dal-r markers were greatly reduced homospecifically. Heterospecifically, the ery-r marker retained all its activity and it could be shown that the size of the segment carrying this activity was of the order of 10^6 daltons. For most markers the efficiency of transformation for segments of DNA of 10^6 daltons is of the order of 1% or less. The str-r, dal-r, and nov-r markers were all greatly reduced in heterospecific activity, i.e. less than 1% but the ery-r marker transforms with an efficiency of 10-20% (Nickel, 1964).

Sucrose gradient fractionation of DNA after sonication confirms the fact that small segments of H. parainfluenzae DNA can transform heterospecifically for the ery-r marker, and demonstrate that the efficiency of heterospecific transformation for smaller segments carrying ery-r is relatively higher in H. influenzae than homospecifically in H. parinfluenzae.

These data suggest that the small elements of H. parainfluenzae DNA in general do not carry the sites for recombination homospecifically and that recombination sites for ery-r marker are different for H. influenzae. These sites are encompassed in the segment effective for transformation of the ery-r marker into H. influenzae. Since the same size segments do not carry much biological activity for the str-r, nov-r, or dal-r markers one must conclude that these segments do not contain sites required for efficient integration into H. influenzae. The correlation between exclusion of H. parainfluenzae markers by H. influenzae in vivo (as measured by heterospecific transformation) and the effect of endo-R in vitro suggests that endo-R plays a specific role in excluding heterologous DNA.

(f) Effect of endo-R treatment on transforming DNA containing markers derived by mutation and heterospecific transformation

It has been known for some time that markers transformed heterospecifically can be subsequently used to transform the recipient with high efficiency. In addition, we have found that at least some heterospecifically transformed markers extracted from its recipient transform the original donor with relatively low efficiency although in some cases these efficiencies are somewhat higher than that of a mutant marker from the same strain (see Table V). We have examined the effect of endo-R on the transforming ability of markers derived heterospecifically versus markers obtained by mutation and found that endo-R was equally effective on the heterospecifically derived marker as on the mutant. These data are presented in Table V. The data are presented in terms of the percentage of transformants obtained using untreated DNA and show that the restriction site of the donor DNA is lost during heterospecific transformation. This loss can be interpreted in one of two ways; either the restriction sites are eliminated by the transformation, or the DNA, when introduced into the recipient, is modified to take on the restriction properties of the recipient.

2. SUMMARY STATEMENT

The restriction of transforming DNA in heterospecific transformation is a problem created by sequence differences between the donor and the recipient. It is now clear that specific restriction enzymes can play a role in the exclusion of heterospecific DNA. In some cases it can play a role in the enhancement

TABLE V

EFFECT OF endo-R TREATMENT ON TRANSFORMING DNA CONTAINING MARKERS DERIVED BY MUTATION AND HETEROSPECIFIC TRANSFORMATION

DNA	Enzyme endo-R	Recipient*	
		H. para	H. inf.
H. inf. ery-r mutant	H. para	10	<1**
	H. inf.	100	100
H. inf. ery-r trans from H.para	H. para	10	<1
	H. inf.	---	100
H. para ery-r mutant	H. inf.	<1	100
	H. para	100	100
H. para ery-r trans from H.inf.	H. inf.	<1	100
	H. para	100	100
H. para str-r mutant	H. inf.	<1	<1
	H. para	100	100
H. para str-r trans from H.inf.	H. inf.	2	10
	H. para	100	100

* DNAs are used to transform both H. para and H. inf., i.e. homospecifically and heterospecifically.

** % activity remaining at 60 min.

of transformations, e.g. by reducing the size of H. parainfluenzae DNA and thus reducing the level of lethality induced by large segments of this DNA in H. influenzae. One may suspect that successful heterospecific transformation gives rise to hybrids which transform the recipient efficiently because of a reduction in the amount of sequence inhomology between donor and recipient (e.g. see Biswas and Ravis, 1971; Wilson and Young, this Symposium). Sequence inhomology may be reflected as differences in restriction sites, ability to be modified, or structural changes in the DNA.

Another kind of problem is implicated in systems in

which the efficiency of transformation is high, yet specific restriction activities are found in vitro, e.g. the ability of endo-R from H. influenzae to attack the DNAs of H. influenzae Reid and H. aegyptius even though these DNA can transform H. influenzae heterospecifically very effectively. The reciprocal experiments with H. aegyptius endo-R gave the same result and suggested that there is a process controlling the restriction activity of endo-Rs in vivo.

3. REFERENCES

BEATTIE, K.L. & SETLOW, J.K. (1970). J. bact. 104:390.

BENDLER, JOHN (1968). Ph.D. Thesis, Johns Hopkins University, Baltimore, Maryland.

BISWAS, G.D. & RAVIN, A.W. (1971). Mol. Gen. Genet. 111:1.

GOODGAL, S.H. & GROMKOVA, R. (1972). Proc. Nat. Acad. Sci., Wash., in press.

GROMKOVA, R. & GOODGAL, S.H. (1972). J. Bact. 109:987.

KELLY, T.J., JR. & SMITH, H.O. (1970). J. Mol. Biol. 51:393.

LEIDY, G., JAFFEE, I. & ALEXANDER, H.E. (1962). Proc. Soc. Exptl. Biol. Med. 111:725.

LEIDY, G., HAHN, E. & ALEXANDER, H.E. (1956). J. Exptl. Med. 104:305.

NICKEL, LOIS (1964). Thesis, University of Pennsylvania, Philadelphia, Pennsylvania.

RANDOLPH, M.L. & SETLOW, J.K. (1972). J. Bact. 111:186.

SCHAEFFER, P. (1958). Symp. Soc. Exptl. Biol. 12:60.

SMITH, H.O. & WILCOX, K.W. (1970). J. Mol. Biol. 51:379.

STEINHART, W.L. & HERRIOTT, R.M. (1968). J. Bact. 96:1725.

WILSON, G. & YOUNG, F. (1972). This Symposium.

RESTRICTION OF DNA IN HAEMOPHILUS INFLUENZAE

S. W. GLOVER

Genetics Laboratory, Physics Building,
University of Newcastle
Newcastle-upon-Tyne, NE1 7RU, England

ABSTRACT

Each of the main serotypes of *Haemophilus influenzae* exerts type-specific restriction on DNA which can be detected using the *Haemophilus* phages HP1 and S2. Each strain restricts the growth of these phages previously grown on other strains.

Three strains have been examined in detail, R*a*, R*b* and R*d*. A summary of these experiments will be presented and discussed in the light of recent studies on restriction endonucleases isolated from *H. influenzae*.

1. INTRODUCTION

The reason for our interest in *Haemophilus influenzae* stems from a long-term interest in the host-specificity of bacterial DNA.

It has been known for a long time that many species of bacteria possess a mechanism for screening invading DNA molecules.This mechanism is independent of the source of the DNA and rejects those that are detected as 'foreign', that is, of a different host-specificity from that of the recipient strain. The rejection mechanism involves primarily the action of specific deoxyriboendonucleases which cleave the DNA molecules at certain specific sites. They are able to do this as a consequence of recognising the base sequence at these sites. Since these sequences are relatively short they occur in the DNA of the restricting strain itself which must therefore be possessed of a mechanism preventing endonuclease attack upon its own DNA. This second mechanism is the process of DNA modification. Modification frequently consists of the alteration by methylation of a base or bases in these short sequences so that the endonuclease is no longer able to bind to the substrate.

The first of these sequences to be analysed was the one recognised by an endonuclease from *Haemophilus influenzae*. This enzyme has been isolated and purified from *Haemophilus influenzae* Rd. It degraded foreign DNA but not *Haemophilus influenzae* DNA, and the sequence of bases around the cleavage point has been determined. (Smith & Wilcox, 1970; Kelly & Smith, 1970).

By this time we had considerable knowledge of the genetic control of restriction and modification in *Escherichia coli* K12 and related strains. It was possible to build up a picture of the endonucleases and methylases involved in these host-specificity systems. There is good evidence that the endonuclease consists of three different sub-units each coded for by a different gene.

These genes have been given the symbols *hss*, *hsr* and *hsm*, and it is an extension of the three gene hypothesis that the *hss* coded sub-unit plays an essential role in site recognition. The methylase is controlled by two genes *hss* and *hsm* and thus contains only two types of sub-unit. There is evidence that some plasmid controlled host-specificity systems might be somewhat simpler - controlled only by two genes. (See reviews by Arber & Linn, 1969;Boyer, 1971;Meselson, Yuan & Heywood, 1972).To bridge the gap between, on the one hand, the genetic studies in *E. coli* and, on the other hand, the sequence studies in *Haemophilus influenzae,* we began a study of the genetics of restriction and modification of DNA in *Haemophilus*.

2. RESTRICTION AND MODIFICATION OF DNA IN *HAEMOPHILUS INFLUENZAE* R*D*.

The *Haemophilus influenzae* strain R*d* is the one most widely used in transformation studies and was the strain from which the restriction endonuclease had been isolated. It was a basic assumption at the start of these experiments that the endo nuclease R isolated by Smith & Wilcox (1970) was a component of a restriction and modification system. Our objective was to isolate mutants deficient in endonuclease R and/or deficient in the modification that was assumed to be part of this host-specificity system. Fundamentally these experiments failed - in that we did not isolate mutants deficient in endonuclease R, but they did enable us to define, at least in biological terms, a new restriction enzyme and a new modification enzyme.

In the first series of experiments strain R*d* was lysogenised with the *Haemophilus* phage HP1;the lysogenic culture was treated with the mutagen N-methyl-N-nitroso-N'-nitroguanidine (Glover & Piekarowicz,1972)and plated out on Brain-Heart Infusion (BHI)agar to form colonies.The colonies were then replica-plated on to lawns of the sensitive indicator strain R*d*. It was anticipated that lysogenic colonies of the wild-type strain R*d* would produce a zone of lysis on the indicator plates caused by the spontaneous release of phage HP1. Lysogenic colonies of strain Rd in which the modification process was deficient were expected to yield HP1 phage which would be restricted by the indicator strain and thus these colonies would produce a reduced zone of lysis or no lysis at all (see Fig. 1).

It proved relatively easy to obtain colonies which failed to lyse the indicator strain but the reason for this failure was most frequently due to the presence of defective phage (Stuy, 1969) rather than to restriction.

In the second series of experiments exposure of phage HP1 to the mutagen was avoided. Cultures grown from colonies of a mutagenised suspension of Haemophilus influenzae Rd were infected with a predetermined number of HP1 phages and the phage was allowed to grow. After growth a carefully controlled volume of the suspension was plated on the indicator strain Rd. It was anticipated that phage from wild-type Rd cultures would produce plaques on the indicator strain but that phage from modification deficient mutants of Rd would not, or would produce very few plaques (see Fig. 2).

As before it proved rather easy to obtain mutants which appeared, by this test, to be modification deficient. However, further investigation showed that the reason why phage from these mutants produced few plaques on the indicator strain was because there was very little phage in the lysates and not because the phage was restricted by the indicator bacteria.

We were able to show that these mutants, instead of being modification deficient as we had supposed, were able to restrict the infecting phage and thus very few of the infected bacteria produced progeny phage. Since initially the strain had been assumed to be r^+m^+(restriction and modification proficient), we could suppose that the mutants were $r^{+'}$ $m^{+'}$ where prime (') indicates an altered host-specificity. These mutants would have been of great interest since persistent attempts to isolate mutants with altered host-specificity in Escherichia coli had previously failed (Glover, unpublished). However, it was soon discovered that the starting strain was r^- m^- and that the so-called mutants were r^+ $m^+$and that both of these phenotypes are unstable (Fig. 3).

The genetic basis of this instability remains obscure. The most obvious explanation of these observations is that the instability of this system is determined by the alternating locations of a plasmid which can be irreversibly lost.However,treatments which effectively cure Escherichia coli of plasmids failed to have consistent effects on this phenomenon.Attempts to transfer the plasmid from r^+ m^+ strains to r^- m^- 'stable' strains in mixed culture were unsuccessful,although some initial experiments appeared to demonstrate that the plasmid was transferable. These initial experiments were explained when it was found that the 'stable' r^- m^-strains do in fact segregate r^+ m^+strains rarely - with a frequency 10 to 100 times lower than unstable r^-m^-strains.

One other point of interest is that another characteristic of Haemophilus influenzae Rd shows almost identical instability,that is,resistance to phage HP1 (Stuy, 1968).However, these two instabilities, host-specificity and phage resistance, are genetically independent and clearly controlled by different elements.

That the restriction endonuclease,which must be a component of the r_D^+ m_D^+ host-specificity system just described, is not the restriction endonuclease described by Smith & Wilcox (1970)became apparent when it was shown that our r_D^- mutants still contained endo. R (Paul Roy, personal communication).Thus Haemophilus influenzae Rd contains at least two restriction endonucleases.

Results similar in all important respects to these have been obtained by Goodgal (Gromkova & Goodgal, 1972; and personal communication), using Haemophilus phage S2 and the role that restriction endonucleases play in heterospecific transformation has also been investigated by Goodgal (Goodgal - This Symposium). Obviously,the restriction exercised by r_D^+ m_D^+ strains against the DNA of phage from r_D^- m_D^- strains presents a potential problem in any genetic transformation experiments carried out between Rd strains which differ in host-specificity.

3. HOST-SPECIFICITY OF DNA IN HAEMOPHILUS INFLUENZAE RA

Haemophilus influenzae Ra carries two genetically distinct host-specificity systems A1 and A2, each of which is able to restrict Haemophilus phage HP1, and each of which confers a specific modification on phage grown in strain Ra.Among restriction-deficient mutants isolated from strain Ra, seven of the eight possible phenotypes for these two systems were obtained after either one or two mutational steps. The eighth phenotype was not specifically screened for (Piekarowicz & Glover, 1972). Both of these host-specificities,A1 and A2,are genetically stable in their expression (see Table I).

4. HOST-SPECIFICITY OF DNA IN *HAEMOPHILUS INFLUENZAE RB*

The host-specificity of *Haemophilus influenzae Rb* has been investigated following the methods described by Piekarowicz and Glover (1972). Phage HP1*c*1 grown on other *Haemophilus influenzae* strains forms plaques on strain *Rb* with an efficiency of 3×10^{-4} (Table 1). Restriction-deficient mutants have been isolated after mutagenesis and scored for restriction and modification, using phage HP1*c*1. The mutants display a wide range of restriction-deficient phenotypes on which phage HP1*c*1.0 forms plaques with efficiencies that vary from 1.0 to 1×10^{-3}. That restriction-deficient r_B^- mutants can be isolated as a result of a single mutagenic treatment indicates that strain *Rb* possesses only a single host-specificity system, in contrast to strain *Ra* (see above) which has two such systems A1 and A2 (Piekarowicz & Glover, 1972). The modification phenotypes of the mutants so far examined do not display the same variety as the restriction phenotypes. The mutants are either m_B^+ or m_B^-. The $r_B^+ m_B^+$ host-specificity system defined here does not display the instability characteristic of strain *Rd*.

5. HOST-SPECIFICITY OF DNA IN *HAEMOPHILUS INFLUENZAE RC*

Phages HP1 and S2 grown on other strains of *Haemophilus influenzae* does not form plaques on strain *Rc*.The reason for this failure to form plaques has not been investigated. If strain *Rc* is resistant to phage HP1 then sensitive mutants are rare since several thousand colonies have been screened after mutagenesis without yielding a sensitive survivor.

6. HOST-SPECIFICITY OF DNA IN *HAEMOPHILUS INFLUENZAE* *RE* AND *RF*

Phage HP1*c*1 grown on other strains of *Haemophilus influenzae* forms plaques on strains R*e* and R*f* with a very low efficiency (approximately 1 x 10^{-9}). On the other hand, phage HP1 previously grown on strain R*e* or R*f* forms plaques on R*e* and R*f* with an efficiency of 1.0 or very close to 1.0. These strains therefore possess a host-specificity quite distinct from the other *Haemophilus influenzae* strains and it is probable that R*e* and R*f* do not differ from one another in their DNA host-specificities (see Table 1).

The extremely low efficiency of plating of phage HP1.0 on strains R*e* and R*f* very probably indicates the presence of two and possibly even three independent restriction endonucleases.

It has been known for a long time that strain R*d* is the strain most readily transformable by isolated DNA. Other strains have been reported to be poorly transformable or not transformable at all. Whether the reason for this poor transformability is due to problems of competence or whether it is due to restriction of transforming DNA is under investigation.

In summary, these experiments define at least five and possibly seven different restriction endonucleases potentially able to degrade heterologous DNA molecules.

TABLE I

EFFICIENCY OF PLATING OF PHAGE HP1 ON STRAINS OF HAEMOPHILUS INFLUENZAE

Phage grown on *H. influenzae*	Host strains*						
Strains	0	Ra	Rb	Rc	Rd	Re	Rf
HP1 . 0	1.0	5×10^{-6}	3×10^{-4}	n.p.	2×10^{-2}	1×10^{-9}	1×10^{-9}
HP1 . Ra	0.5	1.0	3×10^{-4}	n.p.	2.5×10^{-4}	c 10^{-9}	c 10^{-9}
HP1 . Rb	1.0	5×10^{-6}	1.0	n.p.	1×10^{-3}	c 10^{-9}	c 10^{-9}
HP1 . Rd	1.0	5×10^{-6}	3×10^{-4}	n.p.	1.0	c 10^{-9}	c 10^{-9}
HP1 . Re	1.0	5×10^{-6}	3×10^{-4}	n.p.	1×10^{-3}	1.0	0.5
HP1 . Rf	1.0	5×10^{-6}	3×10^{-4}	n.p.	1×10^{-3}	0.5	1.0

* The strain designated 0 is an $r_D^- m_D^-$ derivative of *Haemophilus influenzae* Rd; the strain designated Rd is an $r_D^+ m_D^+$ derivative of *Haemophilus influenzae* Rd; n.p. indicates that no plaques were obtained from the highest concentration of phage lysate tested (1×10^{10} particles/ml).

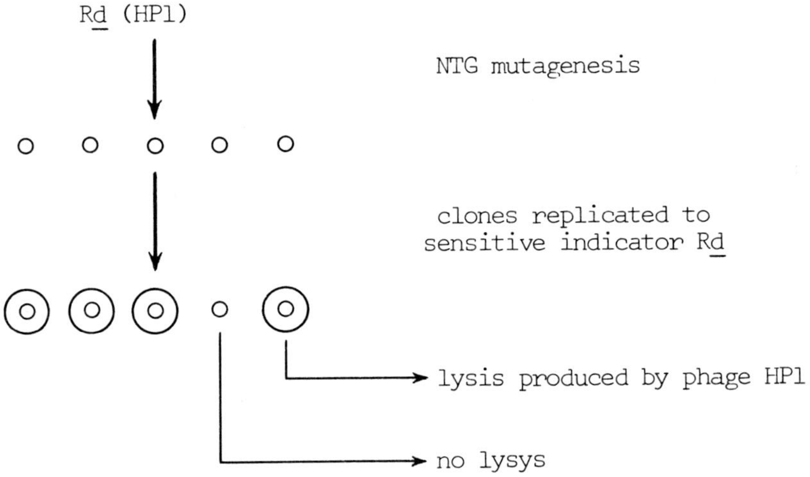

Fig. 1 - Experimental procedure used in initial experiments designed to isolate modification deficient mutants of *Haemophilus influenzae* *Rd*. For further explanation see text.

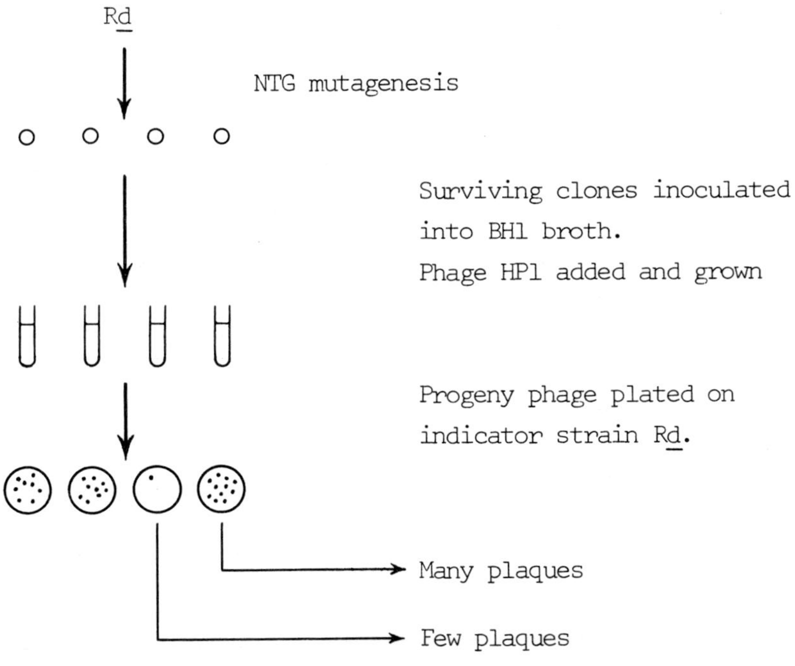

Fig. 2 - Experimental procedure used in further experiments designed to isolate modification deficient mutants of *Haemophilus* influenzae *Rd*.For further explanation see text.

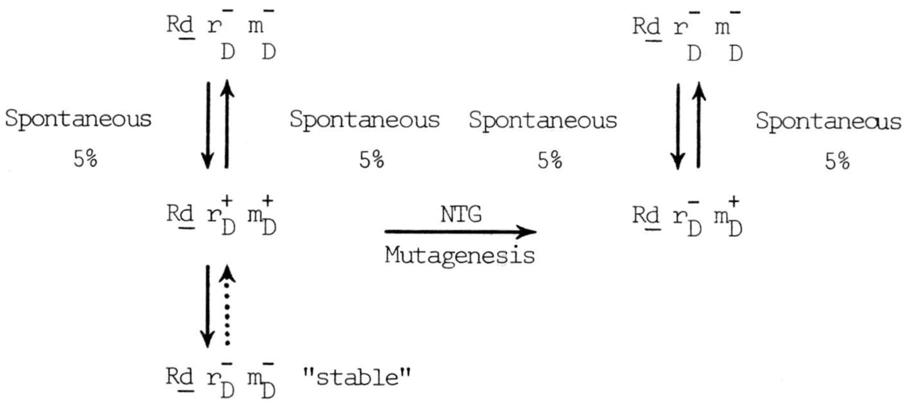

Fig. 3 - Instability of expression of host-specificity type D in Haemophilus influenzae Rd(after Glover & Piekarowicz, 1972).

7. REFERENCES

ARBER, W. & LINN, S. (1969). Ann. Rev. Biochem. 38:467-500.

BOYER, H. W. (1971). Ann. Rev. Microbiol. 25:153-176.

GLOVER, S. W. & PIEKAROWICZ, A. (1972). Biochem. Biophys. Res. Commun. 46:1610-1618.

GROMKOVA, R. & GOODGAL, S. H. (1972). J. Bact. 109:987-992.

KELLY, J. T. & SMITH, H. O. (1970). J. Molec. Biol. 51:393-409.

MESELSON, M., YUAN, R. & HEYWOOD, J. (1972). Ann. Rev. Biochem. (In press).

PIEKAROWICZ, A. & GLOVER, S. W. (1972). Molec. Gen. Genet. 116:11-25.

SMITH, H. O. & WILCOX, K. W. (1970). J. Molec. Biol. 51:379-391.

STUY, J. H. (1968). Biochem. Biophys. Res. Commun. 33:682-687.

STUY, J. H. (1969). Virology, 38:567-572.

INACTIVATION OF TRANSFORMING H. IN

H. KOPECK

M. R. CHEVAI

A. PRUNE

G.BERNA

Laboratoire de Génétique
Institut de Botanique,Strasbourg,
and
Laboratoire de Génétique Moléculaire,
Institut de Biologie Moléculaire,
Paris, France

ABSTRACT

The inactivating effect of four different endonucleases on transforming activity of Haemophilus influenzae DNA was investigated.

Results obtained by viscosimetry,analytical band centrifugation analysis in neutral and alkaline solvents and tests of residual transforming activities have demonstrated that the inactivation levels by different enzymes are only function of the total number of phosphodiester bond breaks introduced into the starting native DNA molecules.

1. INTRODUCTION

Inactivation of transforming DNA by enzymic, chemical and physical agents has been extremely useful for understanding the fundamental steps in bacterial transformation, namely DNA [illegible] DNA integration into the host genome (Zamenhof, Alex- [illegible]d Leidy, 1953; Lerman and Tolmach, 1957; Goodgal, 1961; [illegible] 1961;Bodmer and Ganesan, 1964;Bodmer, 1966;Cato and Guild, [illegible]8; Bernardi and Bach, 1968 ; Nicolaieff and Chevallier, 1970; [illegible]horsett and Hutchinson, 1971).

Inactivation by DNases has the advantage over inactiva tion by physical and chemical agents of being due to a single cause, namely phosphodiester bond breakage. It may be wondered, however, whether the degree of inactivation by DNases is only re lated to the number of bonds broken,or if it shows some relation ship with the mechanism of action of the enzyme(single or double breakage of native DNA), or with the nature of the ends formed (3' or 5' phosphate). Results of Bernardi and Bach (1968) ruled out the first possibility,but not the second one. In fact, these authors showed that at comparable levels of degradation of _H.influenzae_ DNA, spleen acid DNase, a 3'-phosphate former, was much more inactivating than either pancreatic DNase or _E. coli_ endonuclease I, two 5'-phosphate formers. Indirect estimates of the total number of bonds broken by spleen and pancreatic DNases suggested that the different inactivation was not due to a larger number of bonds broken by the spleen _versus_ the pancreatic enzyme hinting at a possible role of the nature of the ends. The subsequent report by Laipis,Olivera and Ganesan(1969)that the ratio of total number of bonds broken to the number of double breaks introduced by pancreatic DNase had been grossly overestimated by previous authors suggested to us that the conclusions of Bernardi and Bach (1968), who had used such estimates, might be incorrect. We were therefore encouraged to re-examine the problem of the inactivation of transforming DNA by DNases. While more detailed data will be reported elsewhere,we present here results obtained by degrading _H. influenzae_ transforming DNA by four different nucleases (spleen acid DNase, pancreatic DNase, *E. coli* endonuclease I, and micrococcal nuclease) and by correlating the inactivation of biological activity with the total number of breaks and the number of double breaks introduced by each enzyme. The results obtained show that inactivation of transforming *H. influenzae* DNA by the DNases used is only related to the total number of breaks introduced by the enzymes, independently of the nature of the ends formed. This conclusion is in agreement with the original proposal of Bodmer (1966) for *B. subtilis* DNA inactivation and with the results of Thorsett and Hutchinson(1971) on the *D. pneumoniae* DNA.

2. RESULTS AND DISCUSSION

DNase-treated DNA's were subjected to band sedimentation analysis in neutral and alkaline conditions (Studier, 1965). The sedimentation data were used to calculate double-stranded and single-stranded molecular weights, using the relationships of Studier(1965). Molecular weights were used, in turn, to estimate the number of double breaks and single breaks per million daltons of native DNA (Charlesby, 1954).

(a) Correlation between inactivation and double-stranded molecular weight of degraded transforming DNA

The correlation between double-stranded molecular weights and biological activity of the novobiocin resistance marker is shown in Fig. 1. Two major conclusions can be drawn from these results: 1) The inactivation curves markedly increase in slope below a molecular weight of about 15×10^6. 2) Large

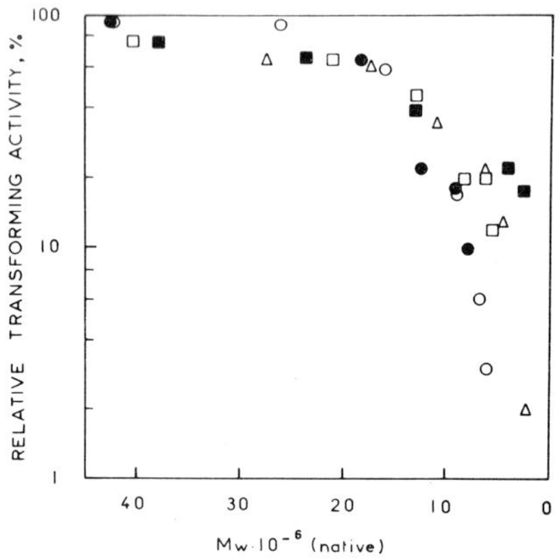

Fig. 1 - Inactivation of novobiocin marker of SN 668 DNA plotted against weight average molecular weights of samples treated by pancreatic DNase in presence of Mg^{++}(○), in presence of Mn^{++} (●), by spleen acid DNase (□), by E.coli endonuclease I(△)and by micrococcal nuclease(■). The residual transforming activities are given relative to that of the stock solution of SN 668 DNA, whose molecular weight was 55 million daltons.The enzyme treatments were performed on DNA dialyzed for 20 hours against the appropriate buffer : 0.15 M sodium acetate buffer-0.01 M EDTA, pH 5.0 for spleen acid DNase;0.02 M Tris-HCl buffer, 0.005 M $MgCl_2$, pH 7.5 for E. coli endonuclease I; 0.1 M borate buffer, $CaCl_2$ - 0.0025 M, pH 8.8 for micrococcal nuclease;0.05 M Tris-HCl, pH 7.5, supplemented with $MgCl_2$ (0.005 M)or $MnCl_2$ (0.00125 M),respectively,for pancreatic DNase.The transformation procedure used was that described by Goodgal and Herriott(1961).DNA concentration in transformation mixture was usually 1 µg/ml.

differences exist in the plots of residual transforming activity *versus* double-stranded molecular weight,as obtained by using different enzymes. On this basis, micrococcal nuclease is the least inactivating enzyme, followed by *E. coli* endonuclease I, spleen acid DNase and pancreatic DNase.

The differences in the inactivation curves shown in Fig. 1 are not related to the nature of the ends formed by different enzymes (3' or 5' phosphate) since 1) dephosphorylation of the 3' ends formed by spleen DNase and of the 5' ends formed by E. coli endonuclease I does not lead to any change in the residual transforming activity (Fig. 2), and 2) inactivation by spleen DNase and micrococcal nuclease, which are both 3'-phosphate formers, is very different.

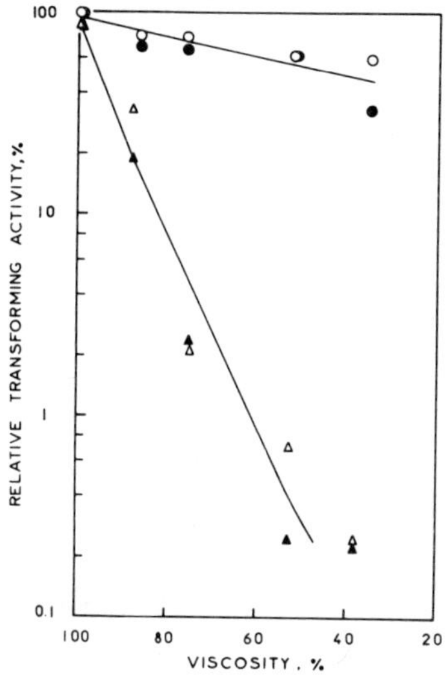

Fig. 2 - Inactivation of novobiocin marker in P^{32} labelled DNA of H. influenzae treated with spleen acid DNase; before dephosphorylation (△) and after dephosphorylation (▲). Inactivation of novobiocin marker in $EVSN_2B$ DNA degraded with E. coli endonuclease I; before (○) and after dephosphorylation (●). Dephosphorylation was carried out according to Weiss, Live and Richardson (1968).

In addition, differences in molecular weight distribution of DNA degraded by different enzymes are not responsible for the differences observed in Fig. 1, as shown by experiments in which DNA samples degraded to comparable double-stranded molecular weights by spleen DNase and micrococcal nuclease were sedimented together in sucrose gradients (Fig. 3). The difference in specific activity in the fragments obtained after digestion with the two enzymes was practically the same for all fragment sizes; as expected, the largest fragments showed the highest specific activity.

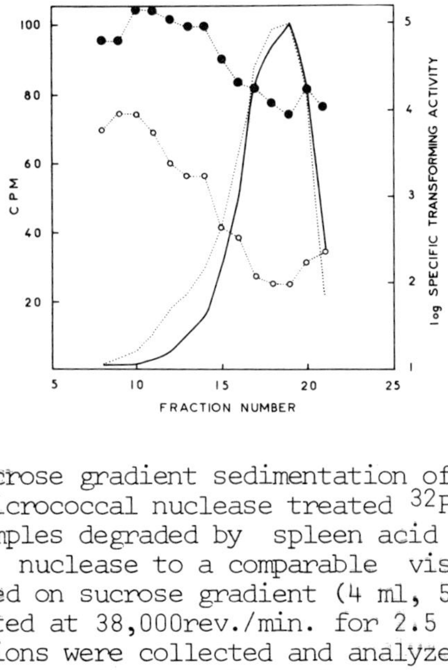

Fig. 3 - Sucrose gradient sedimentation of spleen acid DNase and micrococcal nuclease treated ^{32}P-DNA. ^{32}P-DNA samples degraded by spleen acid DNase and by micrococcal nuclease to a comparable viscosity (50%) were layered on sucrose gradient (4 ml, 5-20% in SSC) and sedimented at 38,000rev./min. for 2.5 hours at 10° C. 21 fractions were collected and analyzed for radioactivity (...... micrococcal nuclease treated sample) (—— spleen acid DNase treated sample), and for biological activity at non saturating DNA concentration (0,01 μg/ml). The biological activity is expressed as the log of specific transforming activity (i.e. log of number of transformants per 10^{-2} μg of DNA).
●.....● micrococcal nuclease treated DNA. O.....O spleen acid DNase treated DNA.

(b) Correlation between inactivation and single-stranded molecular weight of degraded transforming DNA

When residual transforming activity is plotted against the single-stranded molecular weight of H.influenzae DNA degraded by different enzymes, all data practically fit the same line (Fig. 4) showing the existence of a unique relationship between residual activity and single-stranded molecular weight of the fragments, regardless of the enzyme used.

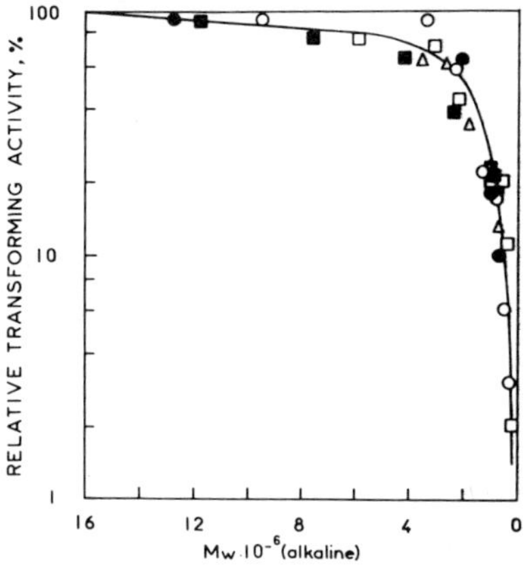

Fig. 4 - Inactivation of novobiocin marker of SN 668 DNA as a function of M_{20,w_A} determined in alcaline conditions. Results were obtained with DNA degraded by pancreatic DNase with Mg^{++} (○) and with Mn^{++} (●), by spleen acid DNase (□), by E. coli endonuclease I (△) and by micrococcal DNase (■).

The slope of the inactivation curve markedly increases below a single-stranded molecular weight of 3-4 x 10^6, a value significantly lower than half the value at which a similar phenomenon takes place for double-stranded molecular weight (Fig. 1). This phenomenon is explained by the presence of nicks in the double-stranded DNA. Fig. 5 shows plots of residual transforming activity against the total number of breaks introduced by different enzymes per one million daltons of double-stranded molecular weight. As expected, points fall on the same line regardless of the enzymes used.

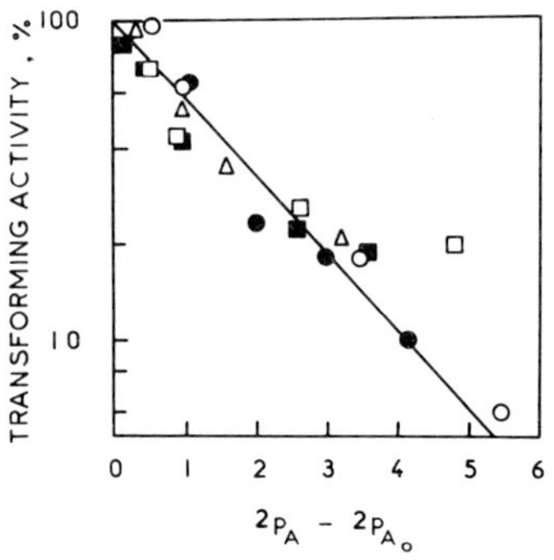

Fig. 5 - Inactivation of novobiocin marker plotted versus total number of breaks introduced by pancreatic DNase with Mg^{++} (○) and with Mn^{++} (●) as activators, by spleen acid DNase (□), by E. coli endonuclease I (△) and by micrococcal nuclease (■). $2pA - 2pA_0$ is the total number of breaks introduced by the enzymes per one million daltons of native DNA.

From the results presented in Figs.4 and 5,we can conclude that the level of residual biological activity in H. influenzae transformation, like in other transforming systems, (B. subtilis, Bodmer, 1966; and D. pneumoniae, Thorsett and Hutchinson, 1971) is a function of the size of single stranded DNA. This could be expected from the findings of Goodgal and Notani (1966)who demonstrated that in H. influenzae transformation only one strand of the DNA taken up by the recipient cells is incorporated into the host genome.

(c) Uptake of degraded DNA

Bernardi and Bach (1968) used competition experiments in order to show that uptake of transforming DNA was not differentially inhibited by spleen acid DNase and pancreatic DNase, and also that uptake was much less impaired than transforming activity, in agreement with Lerman and Tolmach (1957). We have confirmed this latter point directly in the case of spleen DNase degradation by measuring the uptake of tritium labelled DNA (Fig. 6). In another experiment, penetration of spleen DNase-treated DNA before and after dephosphorylation was compared. DNA uptake and transforming activity were not modified by dephosphorylation. This result confirms the idea that 5' or 3' phosphate ends are not involved in inactivation, as already suggested by the previous results.

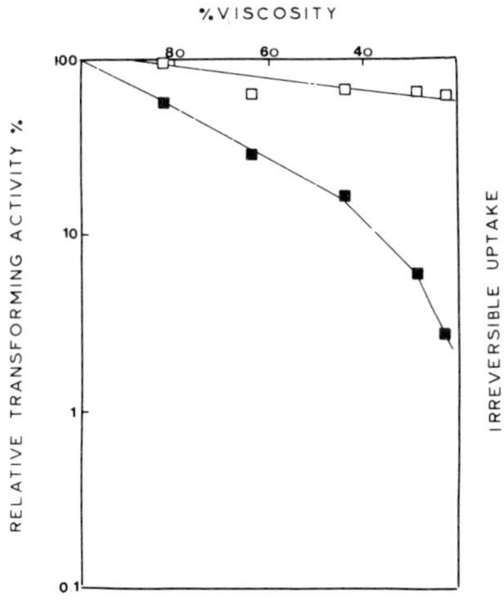

Fig. 6 - Decrease in transforming ability and uptake of ^{3}H-DNA treated by spleen acid DNase as a function of viscosity. Irreversible uptake was measured using the technique of Barnhardt and Herriott (1963). Transforming activity: ■——■ , irreversible uptake: □——□ .

3. REFERENCES

BARNHARDT, B. J. & HERRIOTT, R. M. (1963). Biochim. Biochim. Acta 76:25.

BERNARDI, G. & BACH, M. L. (1968). J. Mol. Biol. 37:87.

BODMER, W. F. (1966). J. Gen. Physiol. 49:233.

BODMER, W. F. & GANESAN, A. T. (1964). Genetics 50:717.

CATO A. & GUILD W. R. (1968). J. Mol. Biol. 37:157.

CHARLESBY, A. (1954). Proc. Roy. Soc. A. 224:120.

GOODGAL, S. H. (1961). J. Gen. Physiol. 45:205.

GOODGAL, S. H. & NOTANI, N. (1966). Fed. Proc. 25:707.

LAIPIS, P. J., OLIVERA, B. M. & GANESAN, A. T. (1969). Proc. Natl. Acad. Sci. 62:289.

LERMAN, L. S. & TOLMACH, L. J. (1957). Biochim. Biophys. Acta 26:68.

NICOLAIEFF, A. & CHEVALLIER, M. R. (1970). Mol. Gen. Genetics 109:269.

STUDIER, F. W. (1965). J. Mol. Biol. 11:173.

STUY, J. H. (1961). PhD. dissertation. Utrecht University

THORSETT, G. O. & HUTCHINSON, F. (1971). Biochim. Biophys. Acta 238:67.

WEISS, B., LIVE, T. R. & RICHARDSON, C. C. (1968). J. Biol. Chem. 243:4556.

ZAMENHOF, S., ALEXANDER, H. E. & LEIDY, G. (1953). J. Exp. Med. 98:373.

DNA-MEMBRANE INTERACTIONS IN THE *B. SUBTILIS* TRANSFORMATION SYSTEM[1,2]

EUGENE W. NESTER
DOUGLAS C. DOOLEY*

Department of Microbiology, School of Medicine
University of Washington, Seattle, Washington 98195

ABSTRACT

The *Bacillus subtilis* transformation system has been studied with regard to the association of the transforming DNA and the membrane of the recipient cell. The DNA-membrane complex was isolated on gradients of Renografin, an X-ray contrast material. Within minutes after uptake, DNA binds to the cytoplasmic membrane where it displays a greater resistance than the recipient genome-membrane complex to dissociation by shearing forces. Ten minutes after uptake, a portion of the donor DNA is released from the membrane. This material is unintegrated DNA which is incapable of transforming other cells. Recombinant or integrated DNA is enriched up to 1.7 fold in the membrane relative to the amount of recipient DNA associated with the membrane. This enrichment lasts at least 30 minutes after termination of DNA uptake and probably much longer. The data suggest that transforming DNA may be integrated into the recipient genome on, or close to the cytoplasmic membrane.

[1]This investigation was supported by National Science Foundation Grant number GB 25167 and National Institute of Health Grant number 5T01 GM005 11-12.

[2]This work was taken from a thesis by Douglas Dooley in partial fulfillment of the Ph.D. degree.

*Present address: Worcester Foundation for Experimental Biology, Schrewsbury, Massachusetts 01545.

1. INTRODUCTION

A number of studies over the past several years have suggested an association between cellular DNA and the cytoplasmic membrane. With regards to the association between added transforming DNA and the membrane in Bacillus subtilis, Wolstenholme, Vermeulen and Venema (1966) found an increase in the number of membranous bodies connecting the nuclear region and the cytoplasmic membrane in cells in competent cultures. Autoradiography with tritiated DNA demonstrated that transforming DNA was associated with these bodies. Using a similar approach, Javor and Tomasz (1968) found 74% of the silver grains of tritiated DNA lying over the tip of the cell and 18% at the exact middle after 90 minute exposure to DNA. Young (1967) observed that after a 30 minute exposure to tritiated DNA, 88% of the DN'ase resistant radioactivity was near the cell wall. Landman and Knott (1968) reported experiments which suggested that DN'ase resistant transforming DNA can be bound outside the cell membrane in the periplasmic space.

For the past year, our laboratory has been interested in following the fate of transforming DNA with regard to its association with the cytoplasmic membrane. The results of these studies will form the basis of this paper.

2. RESULTS

We first determined which of several methods for demonstrating DNA-membrane complexes would be most satisfactory for studying DNA-membrane complexes in B. subtilis. Initially, the technique reported by Tremblay, Daniels and Schaechter (1969) was used. In this technique, protoplasts, formed by the action of lysozyme, are lysed directly on a sucrose gradient with sarkosyl and the lysate centrifuged.

In the presence of magnesium,sarkosyl crystallizes and the DNA attaches.This technique is convenient and rapid.However, a problem arises with its use in Bacillus subtilis. The fraction of the genome recovered in the M-band is a function of the concentration of lysozyme in the lysate.

Strain 168 was labeled with tritiated thymidine. The cells were converted to protoplasts by incubation with 1 μg/ml lysozyme for 60 min in sucrose-Tris-Mg-KCl buffer. The protoplasts were distributed into tubes containing different amounts of lysozyme; sarkosyl was added and the lysates sedimented to obtain the M-band. The results show that the amount of DNA which is found in the M-band is a function of the amount of lysozyme present (Fig. 1).

This lysozyme effect in M-bands may be unique to B. subtilis, for similar results were not obtained with E. coli or B. megaterium.

Recently,an alternative method was developed specifically for purifying the DNA-membrane complex from B. subtilis,and this method was employed for the rest of the studies.Protoplasts are prepared by treatment with 50 μg/ml lysozyme for 30-60 min at 37C and then opened with 0.01% Brij, a non-ionic detergent. The lysate is sheared and then centrifuged on a 5-50% Renografin gradient on a 70% sucrose shelf for 4.5 hr. At this time, the density of the membrane complex has equilibrated with the density of the Renografin.

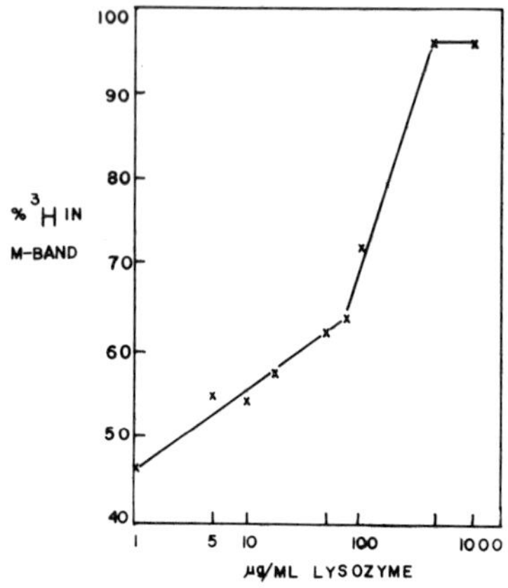

Fig. 1 - DNA Content of the m-band as a Function of the Lysozyme Concentration. Strain 168 was labeled for two generations with 0.66uCi/ml ^{3}H-thymidine in Spizizen's minimal medium supplemented with 0.1% casein hydrolysate + 20 μg/ml tryptophan and 100 μg/ml 2'-deoxyadenosine. Cells were washed twice in 13.5% sucrose-0.01 M Tris pH 7.0, 0.01 M MgAcetate, 0.1 M KCl buffer and resuspended in the same buffer at a Klett reading of 80. The cells were converted to protoplasts with 1 μg/ml lysozyme at 44 C for 60 min. Protoplasts were centrifuged, resuspended in fresh 13.5% sucrose - Tris.MgAcetate,KCl buffer, and distributed into tubes containing different amounts of lysozyme (abscissa). Sarkosyl(0.1% final concentration) was added and the lysates sedimented through sucrose gradients to obtain the m-band.

If whole cells with their proteins and DNA labels are centrifuged, virtually all of both labels are at the bottom of the gradient (Fig. 2a). If the cells are gently lysed, and the lysate centrifuged, virtually all of the DNA sediments with the membrane (Fig. 2b). If the cells are lysed and the lysate subject to shear, then the DNA is sheared off the membrane fraction and is found in the "free" DNA fraction (Fig. 3b). Greater shear results in a greater amount of DNA being removed from the membrane (Fig. 3a). It is our strong feeling that the complex is not

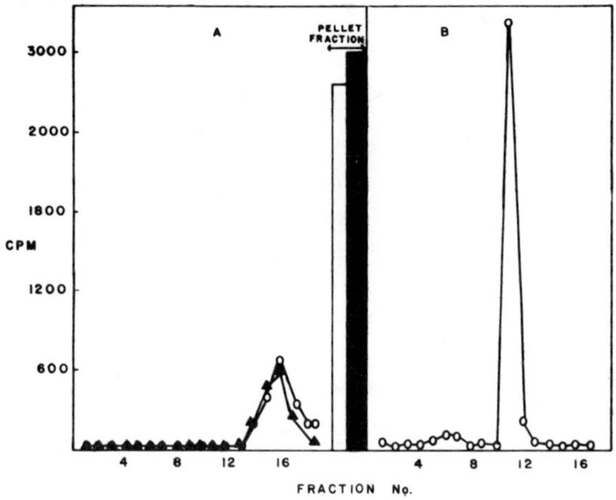

Fig. 2 - Distribution of DNA and Protein on Renografin membrane gradients. Strain 168 (trypC$_2^-$), grown as in Fig. 1, was labeled for 1.5 generations with 0.2 uCi/ml ^{14}C-leucine and 5uCi/ml ^{3}H-thymidine(50 µg/ml 2'-deoxyadenosine). (A) 1 x 10^8 intact cells were layered on a 5-50% renografin gradient. (B) Cells were prepared with a minimum of shear for isolation of the DNA-Membrane Complex on a 5-50% renografin gradient as described in the text. 0—0 ^{3}H-thymidine ▲—▲ ^{14}C-leucine.

an artifact of the preparation and all efforts to demonstrate that the DNA is trapped inside a vesicle have been negative. A large number of DNA preparations from different organisms in both double stranded and single stranded forms did not associate with the membrane when the lysate was centrifuged in Renografin.

The chemical composition of the DNA-membrane complex in <u>B. subtilis</u> is shown in Table I. We have not excluded the possibility that fragments of the cell wall are also present. However, it is possible to isolate the DNA-membrane complex from protoplasts.

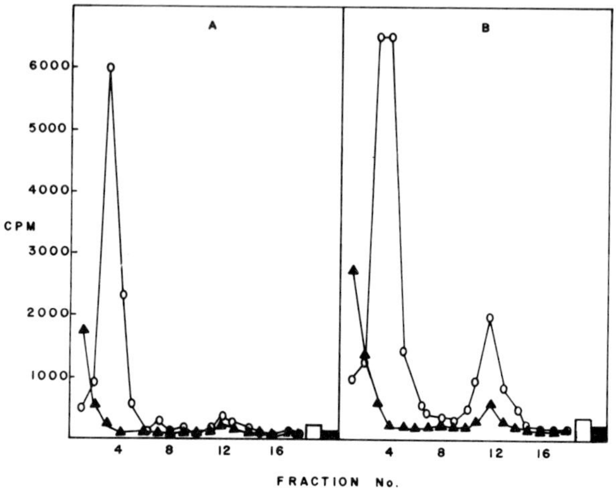

Fig. 3 - Strain 168 was grown and labeled as in Fig. 2. Cells were prepared for isolation of the DNA-Membrane Complex on a 5-50% Renografin gradient as described in the text with the following modifications. (A) Lysate of 1 x 10^8 cells was sheared by 4 rapid passages through a 20 gauge needle. (B) Lysate of 1 x 10^8 cells was sheared for 15 sec. on a Vortex mixer. 0—0 ^{3}H-thymidine ▲—▲ ^{14}C-leucine.

From these experiments, we conclude that by opening the cell gently, it is possible to isolate virtually all of the cellular DNA associated with a membrane complex. Further, it appears that it should be possible to study the strength and nature of this association by determining the ability of shearing forces or chemicals to disrupt this complex.

Is there any difference in the strength of association between the DNA and membrane in competent and non-competent cells? Recall that in competent cultures of B. subtilis generally no more than 10% of the cells are capable of being transformed. Therefore, it was necessary to physically separate competent from non-competent cells. Purified populations of competent and

TABLE I

CHEMICAL COMPOSITION OF DNA-MEMBRANE COMPLEX[1]

Material	Recovered Activity in Complex (%)
Protein	5 - 15
Phospholipid	24 - 26
DNA	30 - 90
Stable RNA	<1
Pulse labeled RNA	4 - 11

(1) Ivarie, R. D. and Pene, J. J. (1970). J. Bact. 104:839.

non-competent cells were obtained by centrifuging a competent population on Renografin.The cells were washed,lysed and analyzed for the amount of free and membrane associated DNA.In this experiment, the competent and non-competent lysates were subjected to 0.01% Brij,a non-ionic detergent.The results indicate that there is no difference between the competent and non-competent cells. Although there is a marked difference in the total counts present in the competent and non-competent fractions,the ratio of counts in the free and membrane-associated DNA are identical (Fig. 4). The amount of DNA in both fractions is approximately the same.In other experiments, the lysates from competent and non-competent cells were subjected to various degrees of shear on a Vortex mixer.In all cases, the percent of DNA associated with the membrane was the same for competent and non-competent cells. Since there is no apparent difference in the degree of association of the DNA with the membrane, all further experiments were done on un-

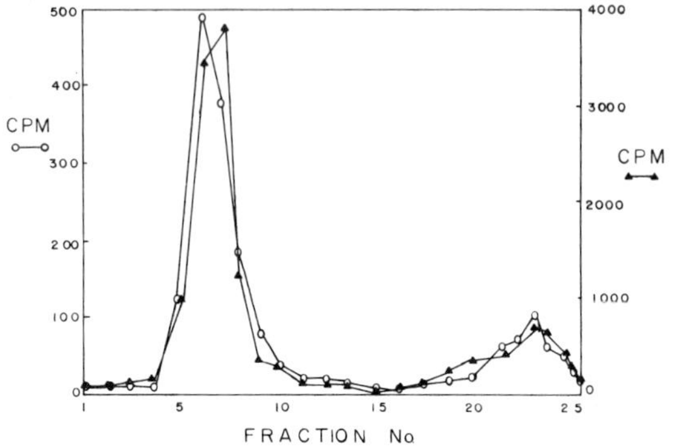

Fig. 4 - Sedimentation of the DNA-membrane complex from purified competent and noncompetent cells. Strain SB25 ($his2^-$, $trypC_2^-$) was brought to maximum competence in MM2 in the presence of 0.5 uCi/ml ^{3}H-thymidine and 20 µg/ml 2'-deoxyadenosine. After transformation with SB32 ($his2^-$) DNA, the culture was separated into competent and noncompetent fractions on 27-35% Renografin gradients. The heavy and light fractions were washed and resuspended in 0.01M NaAcetate, 0.08M K_2HPO_4, 0.04 M KH_2PO_4 buffer pH 7.4 to a Klett reading of 60 and lysed with 100 µg/ml lysozyme. Resulting lysates were sedimented through 5-50% renografin gradients. ▲—▲ Noncompetent cell lysate made 0.01% BRIJ and sheared 15 sec. O—O Competent cell lysate made 0.01% BRIJ and sheared 15 sec.

fractionated cells.

Most of the data that will be presented in this paper are based on radioactivity measurements of high molecular weight DNA. All data are consistent with the notion, however, that labeled DNA is incorporated in macromolecular form and is not broken down to low molecular weight nucleotides and then incorporated. The data which support this conclusion are: (1) label and transformant count decreased to same degree by EDTA, DN'ase, competitor DNA, (2) kinetics of transformation and uptake of label are identical, (3) data of other investigators.

The initial experiment on the fate of transforming DNA was to study the association between the membrane and DNA at various times after the addition of DNA. The DNA of recipient cells was labeled with ^{14}C and the cells were exposed to ^{3}H DNA. Samples were removed at various times to DN'ase for 2 min and then processed. The data are shown in Figure 5. Five min after DNA addition, more than 90% of the transforming DNA is recovered in the

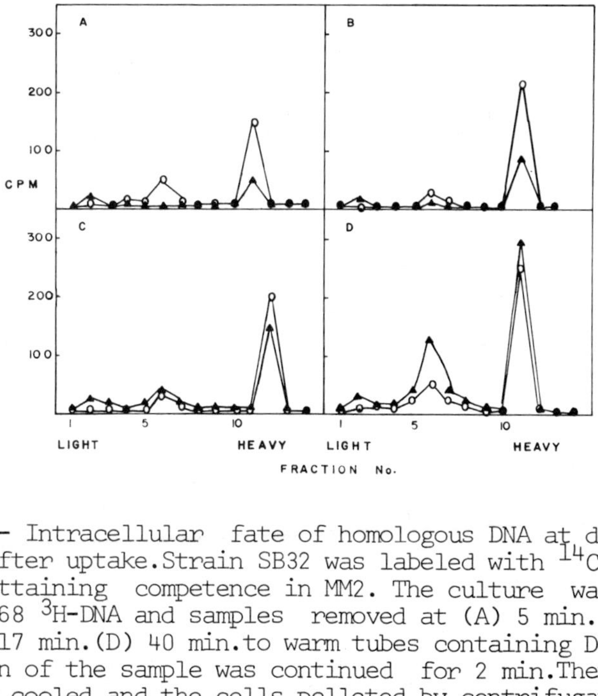

Fig. 5 - Intracellular fate of homologous DNA at different times after uptake. Strain SB32 was labeled with ^{14}C-thymine while attaining competence in MM2. The culture was made 2 mg/ml 168 ^{3}H-DNA and samples removed at (A) 5 min. (B) 10 min. (C) 17 min. (D) 40 min. to warm tubes containing DNase. Incubation of the sample was continued for 2 min. Then it was quickly cooled and the cells pelleted by centrifugation. The pellet was resuspended in buffer containing 10% glycerol and quick frozen in a dry ice-ethanol bath. When sampling has been completed, all were processed together for analysis on 5-50% renografin gradients. No shear was applied during preparation of the lysates and 75% of the recipient cell DNA was membrane bound. ▲—▲ Donor ^{3}H DNA 0——0 Recipient ^{14}C DNA.

membrane complex although only 65% of the recipient DNA is bound to the membrane. After 17 minutes, free DNA is apparent, and as incubation continues, the amount of free as well as membrane bound DNA increases.

The changing distribution of radioactivity in donor DNA with time plotted in Figure 6. Initially, practically all of the

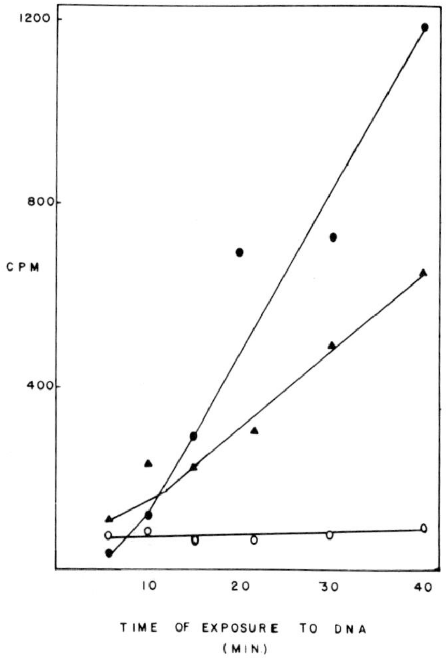

Fig. 6 - Distribution of donor radioactivity during continuous uptake of homologous DNA. Strain SB32 was transformed with 168 ^{3}H DNA as in Fig. 5 and sampled at the times indicated. The lysates were sheared for 3 seconds and sedimented through 5-50% Renografin gradients. Approximately 60% of the recipient DNA was membrane-associated. The proportion of membrane-associated recipient ^{14}C DNA in each sample was used as an internal control to correct ^{3}H values for differences in shear between samples. O——O Low M.W. donor; ●——● High M.W. donor ^{3}H free of the membrane; ▲——▲ Membrane-associated donor ^{3}H.

DNA is membrane bound. With continued incubation, the amount of DNA in the membrane fraction increases; however, the amount of donor material free of the membrane increases at an even faster rate. The amount of low molecular weight material remains almost constant throughout the period of DNA uptake.

One plausible explanation for these results is that the DNA is processed in an assembly line type of processing. In this explanation, each DNA molecule taken up by the cell is membrane bound for 5-10 minutes, after which time two-thirds of each molecule is released from the membrane, either because it is sheared off during the preparation and manipulation of the lysate, or because nucleases promote its release.

To gain data as to whether this explanation is correct, competent cells of 168 were pulsed with tritiated DNA for 10 min and all further uptake was terminated by the addition of DN'ase. Incubation was continued, and post-uptake events were examined by analyzing samples for their DNA distribution at various times. Two minutes after the pulse of DNA, 70% of the high M.W. DNA was membrane associated under conditions which left only 45% of the recipient DNA membrane bound. Seven minutes after the pulse, there were equal amounts of tritium in the membrane and free DNA fractions and after 32-40 minutes, less than one-third of the donor DNA was membrane bound (Fig. 7).

These data are consistent with the notion that DNA undergoes a type of intracellular processing, such that there is a transition from a state of strong membrane association to a state of relatively weak association. This transition can be seen as a loss of counts from the membrane to the free DNA fraction.

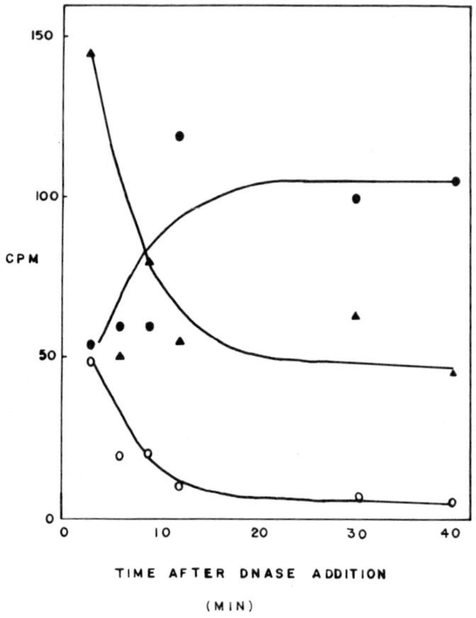

Fig. 7 - Distribution of homologous donor radioactivity after termination of brief exposure to DNA. Competent ^{14}C Thymine labeled cells were transformed for 10 minutes with 168 ^{3}H DNA. At that point, uptake was terminated with DNase and the culture made 50 µg/ml thymine. Samples were removed at the indicated times and processed as in Fig. 5, the lysates receiving 3 seconds shear before sedimentation through 5-50% Renografin gradients. Approximately 50% of the recipient DNA was membrane bound. Data were normalized as in Fig. 6. O——O Low M.W. donor ^{3}H; ●——● Free, high M.W. donor ^{3}H ; ▲—▲ Membrane bound donor ^{3}H.

One of the most significant features of these studies is the observation that the majority of the transforming activity is associated with the membrane. This association is maintained for at least 75 minutes (Fig. 8). In this experiment, a histidine requiring strain was transformed with DNA from a tryptophan auxotroph linked to the his locus. The DNA was in contact 40 minutes. Thus, both recombinant and donor activity could be followed in

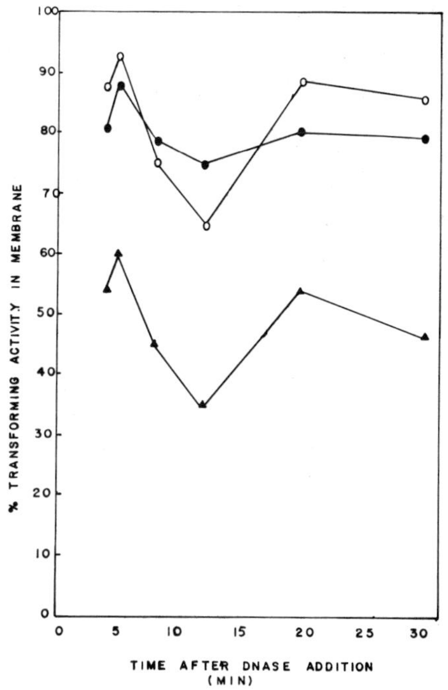

Fig. 8 - Distribution of transforming activity of donor, recombinant and recipient DNA on Renografin membrane gradients. Strain 168 was transformed with SB32 DNA for 44 minutes before termination with DNase. An equal volume of fresh warm medium was added and samples removed at the indicated time for quick-freeze. Samples were prepared for 5-50% Renografin gradients with 3 sec. shear applied to the lysate. Fractions were assayed for biological activity of DNA by transformation. Approximately 50% of the recipient DNA transforming activity was membrane-bound. ●——● Donor marker transforming activity; O——O Recombinant marker transforming activity; ▲—▲ Recipient marker transforming activity.

the reisolated DNA. The donor as well as recombinant activity is enriched in the membrane from 1.5-1.7 fold.

Not only is the association between the donor DNA-membrane complex stronger than the recipient DNA-membrane complex, but its properties are somewhat different (Table II). ^{14}C thymine labeled cells were exposed to tritiated DNA for 12-15 minutes and DN'ase was added. By this time, less than 50% of the donor DNA was integrated. It is clear that without treatment, the donor DNA is enriched in the membrane. Membrane-bound donor DNA exhibits a greater sensitivity to two non-ionic detergents than does the recipient-DNA membrane complex. Further, high temperatures also seem to dissociate the donor DNA-membrane more readily than the recipient DNA-membrane complex. 1 M NaCl dissociates both complexes to the same degree.

TABLE II

CHARACTERIZATION OF TRANSFORMING DNA-MEMBRANE COMPLEX

Treatment	A % ^{3}H in membrane complex	B % ^{14}C in membrane complex	A/B Fold-enrichment of donor DNA in membrane complex
control	67	47	1.43
heat 66 C, 5 min.	57	51	1.12
lysate made 1 M NaCl	48	33	1.46
lysate made 0.5% Nonidet	67	61	1.10
lysate made 0.5% Brij	47	45	1.04

From the foregoing experiments, we conclude that during the 5-10 minute interval between DNA uptake and integration, the vast majority of donor DNA is bound to the membrane. The donor DNA is bound in a way that differs from the binding of the recipient genome: (1) it is more difficult to shear newly absorbed donor DNA than recipient DNA from the membrane and (2) newly absorbed DNA is preferentially released from the membrane by elevated temperatures or non-ionic detergents. A large portion of the donor DNA is released from the membrane after 10 minutes of DNA uptake. The release of donor DNA is detectable approximately at the same time that recombinant DNA can first be detected.

Having defined the kinetics of DNA processing shortly after uptake,the study was extended to longer periods of time in order to investigate events occurring during the reinitiation of competent cell growth and DNA synthesis.A strain was transformed for 50 minutes, DN'ase was added and the cells were resuspended in a medium which only permitted growth of transformants.Samples were taken for the next seven hours incubation, and sedimented through Renografin gradients (Fig. 9). It is evident that there is a loss of ethanol precipitable radioactivity from both the free DNA and membrane fractions in the first hour, after which there is very little change in the distribution of label. Under the conditions of incubation in this experiment, transformants begin to multiply at 4.2 hr. Therefore, as a transformant loses its competence, initiates DNA synthesis and begins to grow, the strength of the DNA-membrane association does not appear to change detectably. This suggests that the integrated fragment of DNA, firmly associated with the membrane soon after integration, is still closely associated with the membrane after several rounds of DNA synthesis and cell replication.

If binding to the membrane is an obligatory step in the uptake of DNA by a cell, heterologous DNA which is taken up into the cell should also be bound to the membrane.To study this possibility, tritiated DNA isolated from cells of E. coli was added to competent cells of B. subtilis. Twenty minutes after DNA addition, DN'ase was added and the culture incubated. At various

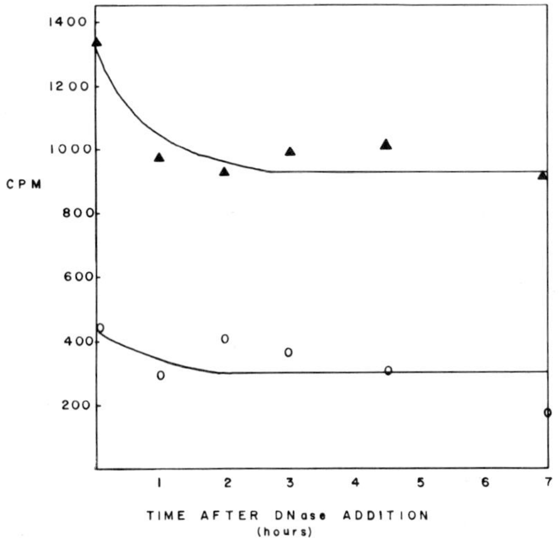

Fig. 9 - Intracellular location of homologous donor DNA during outgrowth from the competent state. Cells were transformed for 50 minutes as in Fig. 5.Then uptake was terminated with DNase, cells were centrifuged,resuspended in fresh medium,and aliquots were removed at the indicated times to be frozen until sampling was completed. Lysates were sheared for 8 seconds and applied to 5-50% Renografin gradients.Twenty per cent of the recipient DNA was membrane-bound. Transformant viable count began to increase 4-5 hours after DNase addition. ▲—▲ Free donor ^{3}H; 0——0 Membrane bound donor ^{3}H.

times, samples were removed and centrifuged. The data of the earliest time,2 minutes after DN'ase addition,are shown in Figure 10.It is clear that the E. coli DNA associates with the membrane. Second, 40% of the donor radioactivity is present at the top of the gradient, indicating extensive hydrolysis of the DNA.In comparison,after 20 minutes exposure to homologous DNA, only 10% of the donor radioactivity is found in low molecular weight fractions. Furthermore, the profile of the free donor DNA is different from the free recipient DNA or that seen with B. subtilis

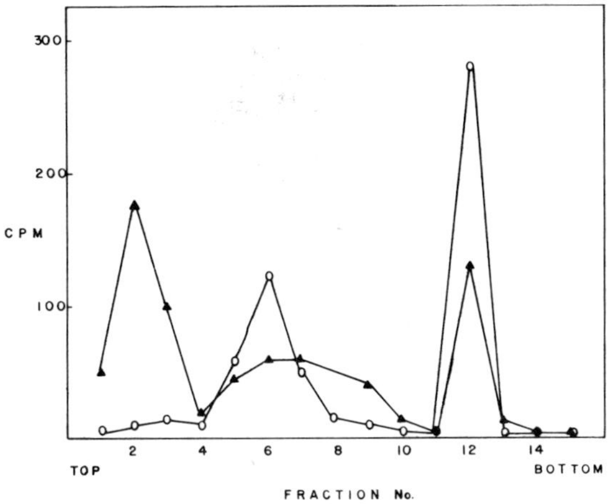

Fig. 10 - Intracellular fate of heterologous DNA at 2 min after termination of uptake. ^{14}C Thymine labeled cells were exposed to 1 µg/ml ^{3}H DNA from E. coli. After 20 min, uptake was terminated by addition of DNase and the culture made 50 µg/ml thymine. Cells were removed at 2 min and treated as in Fig. 5, with three seconds shear applied to the lysates. Fifty per cent of the recipient DNA was membrane-bound. 0——0 Recipient ^{14}C DNA; ▲—▲ Donor ^{3}H DNA.

donor DNA. One reasonable explanation is that the E. coli DNA is undergoing extensive degradation, and perhaps a portion of the free DNA is denatured. Single stranded DNA sediments more rapidly than double stranded DNA on Renografin gradients. Data on additional points are presented in Figure 11.

These data reveal in different form differences between the processing of homologous and heterologous DNA. With E. coli DNA as the donor, there is a continual loss of precipitable radioactivity from all three fractions for the entire 40 minutes (Fig. 11). The differences in processing are also reflected in the recovery of donor radioactivity. While the recovery of homologous donor radioactivity decreases only 40-50% within 1-7 hours, 80%

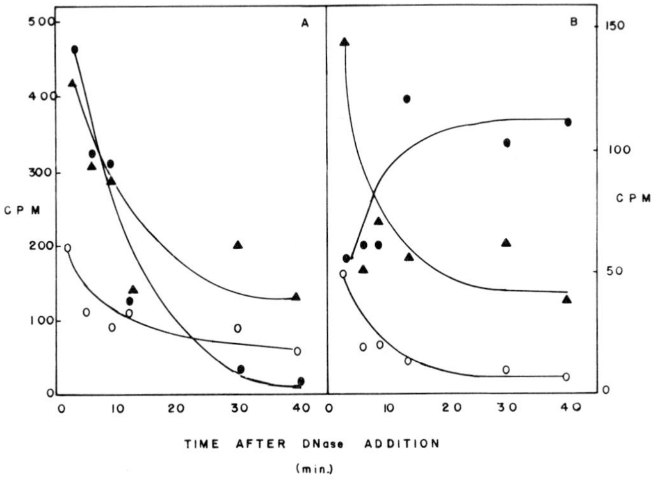

Fig. 11 - Distribution of heterologous and homologous donor DNA radioactivity after the termination of uptake. (A) Labeled cells were exposed to E. coli DNA as in Fig. 10 and samples were taken at the times indicated after DNase addition. The data were normalized as in Fig. 6. (B) Competent ^{14}C-Thymine labeled cells were transformed for 10 min with 168 ^{3}H-DNA. At that point uptake was terminated with DNase and the culture was made 50 µg/ml thymine. Samples were removed at the indicated times and processed as in Fig. 5 with the lysate receiving 3 sec. shear. Approximately 50% of the recipient DNA was membrane bound. Data was normalized in Fig. 6. O——O Low M.W. donor ^{3}H; ●——● Free high M.W. donor ^{3}H; ▲——▲ Membrane-bound donor ^{3}H.

of the heterologous radioactivity becomes ethanol soluble within 40 min (Fig. 12). The decreased recovery indicates that the fragments either become ethanol soluble or are washed or released from the cell.

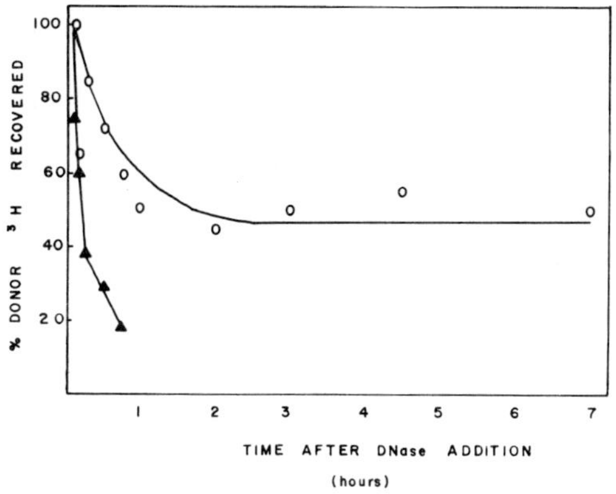

Fig. 12 - Recovery of donor radioactivity with time after termination of uptake. Competent strain 168 was transformed for 10 to 50 min with 168 ^{3}H-DNA or 20 min with E. coli ^{3}H-DNA, and then treated with DNase. Cells transformed for 50 min were treated as in Fig. 9. 0——0 Per cent original homologous donor radioactivity recovered from 5-50% Renografin gradients; ▲—▲ Per cent original heterologous donor radioactivity recovered from 5-50% Renografin gradients.

3. SUMMARY

In summary, within 5-10 minutes after addition, all transforming DNA becomes associated with the cell membrane. This initial association is much stronger than the association of the recipient cell DNA with the membrane, but after about 10 minutes, over half of the donor DNA becomes very sensitive to shear and is readily removed from the membrane. The loci which are integrated, however, maintain this strong association for at least several hours after they have been integrated. Heterologous DNA also binds to the membrane but is very quickly degraded to oligonucleotides which are not alcohol precipitable. The observation that both donor and recipient DNA is enriched in the membrane is consistent with the possibility that integration occurs on or very close to the membrane.

4. REFERENCES

JAVOR, G. & TOMASZ, A. (1968). Proc. Natl. Acad. Sci. U.S. 60:1216.

LANDMAN, O. & KNOTT, R. (1968). Bacteriol. Proc., p.59.

TREMBLAY, G., DANIELS, M. & SCHAECHTER, M. (1969). J. Mol. Biol. 40:65.

WOLSTENHOLME, D., VERMEULEN, C. & VENEMA, G. (1966). J. Bacteriol. 92:1111.

YOUNG, F. (1967). Nature 213:773.

EVIDENCE FOR A NON-COVALENTLY BONDED INTERMEDIATE IN RECOMBINATION DURING TRANSFORMATION OF *BACILLUS SUBTILIS*

F. ARWERT

G. VENEMA

Institute of Genetics, P.O. Box 14,
Haren (Gn), The Netherlands

ABSTRACT

Recombinant type and donor marker activity in extracts of transforming *Bacillus subtilis* cells increase upon incubation with the enzyme polynucleotide ligase. Joint incubation of the re-extracted DNA with both DNA polymerase (Kornberg enzyme)and DNA ligase does not increase further recombinant type and donor marker activity. From these results a non-covalently bonded donor-recipient DNA complex is postulated.

1. INTRODUCTION

There is at present essential agreement that recombination in transformation occurs by breakage and reunion (Bodmer, 1966; Fox, 1966; Notani and Goodgal, 1966).Results obtained from CsCl density gradient centrifugation of both native and denatured recombinant structures have indicated that single-stranded donor fragments are linearly inserted into recipient DNA molecules. Intermediate steps in the integration process,however,are mostly of speculative nature.From the fact that the ultimate product of recombination is a heteroduplex DNA molecule,one is led to postulate an intermediate in which the donor and recipient moieties are held together by non-covalent bonds. Such an intermediate will have single strand interruptions, which, depending on the recombination mechanism,may be either (1) lacking the phosphodiester binding between the donor and recipient strand (nicks) or (2) missing a number of nucleotides adjacent to the inserted segment of donor DNA (gaps),at one or both sides of the inserted single strand. In this paper we present evidence that a structure corresponding to (1) is present in extracts from transforming *Bacillus subtilis* cells.

2. MATERIALS AND METHODS

Strains

Strains used for transformation were recipient 1: *trpC*$_2$ (= *ind*$_{168}$) and recipient 2 : *trpC*$_2$ *tyr*$_1$. The linked markers *trpC*$_2$ *tyr*$_1$ show 70% joint transfer in transformation. Donor DNA was isolated from strain *tyr*$_1$. All strains used were derivatives from *B. subtilis* strain 168.

Transformation

Recipient cells were made competent by a modification of the method of Bott and Wilson (1967). Cells were grown overnight,sedimented and resuspended in glucose-minimal medium (Spizizen, 1958), supplemented with $MgSO_4$ (5×10^{-3} M), 50 µg/ml of each of competence stimulating amino acids(Wilson and Bott,1968): histidine, tryptophane, arginine, valine, lysine, threonine, glycine, aspartic acid and methionine; tyrosine (20 μg/ml) was added when recipient 2 cells (trpC_2 tyr$_1$) were made competent. After resuspension the optical density at 450mμ (OD_{450}) amounted to 0.5, as measured in a Zeiss PMQ II spectrophotometer.The culture was aerated at 37°C for 3.5hr,diluted twofold with glucose-minimal medium and left for an additional 90 minutes with vigorous aeration.During this period the temperature was shifted slowly to 30°C(1°C/13 min.).Transformation frequencies were measured as previously described (Venema, Pritchard and Venema-Schröder, 1965 a).

DNA preparations

Donor DNA was prepared by a modification of Kirby's procedure (1957) as described by Venema *et al.* (1965 a), with an additional phenol extraction after digestion with RNase. Donor DNA had an average double strand molecular weight of 5-8x10^7, determined by analytical boundary sedimentation.Re-extracted DNA was purified less extensively. After phenol extraction the DNA was precipitated with ethoxy-ethanol only once.

Experimental procedure

Competent trpC_2 (recipient 1) cultures were exposed to tyr$_1$ DNA (1 μg/ml) at 30°C for 10 minutes. Further uptake of DNA was stopped by the addition of excess calf thymus DNA(100 μg/ml). At various times afterwards, samples of recipient culture were rapidly chilled, washed four times by centrifugation in ice cold Spizizen's minimal medium and resuspended in SSC. Lysates were obtained by addition of lysozyme(0.5 mg/ml)and 15 minutes storage at 4°C. Lysis was completed by warming to 37°C for 4 min. and by the addition of sodium dodecyl sulfate(0.1%).DNA was re-extracted from samples as described under DNA preparations.

Enzymes

DNA ligase, DNA polymerase and exonuclease III,all obtained from *E.coli*,purified by DNA-agarose column chromatography, were generous gifts from Mr. H.L. Heijneker.DNA ligase,DNA polymerase and exonuclease III were free of other nuclease activity. Enzyme treatment of re-extracted DNA (3 µg in 0.1 ml phosphate buffer) was essentially similar to that described for UV-irradiated DNA by Heijneker, Pannekoek, Oosterbaan, Pouwels, Bron, Arwert and Venema (1971).Controls were treated similarly, except that the enzyme solution was replaced by buffer (0.01 potassium phosphate + 0.001 M EDTA, pH = 7.5). The reaction was terminated by the addition of 0.5 ml 0.5 M NaCl and heating to 68°C for 20 minutes.DNA ligase activity was measured routinely by the repair of transforming activity of DNase I(pancreatic DNase)nicked DNA. Nicked DNA was prepared as described by Laipis, Olivera and Ganesan (1969). In our conditions transforming activity of DNase I inactivated DNA (10-20% residual transforming activity) was restored to 80-100% upon incubation with ligase. Exonuclease III activity prevented restoration of transforming activity of nicked DNA by DNA ligase. However, incubation of nicked DNA with DNA ligase,exonuclease III and DNA polymerase restored the transforming activity again to 80-100%, indicating that the polymerase activity dominated the exonuclease activity.

3. RESULTS

$TrpC_2^+$ tyr_1 DNA, final concentration 1 µg/ml, was used

to transform a competent $\overline{trpC_2 tyr_1^+}$ culture at 30°C for 10 minutes. Samples were removed at 2, 5, 10, 20 and 40 minutes after the addition of calf thymus DNA, chilled, washed and DNA was extracted as described in Materials and Methods. Resident marker activity (tyr_1^+), donor marker activity ($trpC_2^+$) and recombinant type activity ($\overline{trpC_2^+ tyr_1^+}$) were assayed in samples incubated with and without DNA ligase, by transformation of $\overline{trpC_2 tyr_1}$ cells. Fig. 1 shows that in the two-minute and five-minute sample recombinant type activity is increased two fold and donor marker activity 1.4fold upon incubation with DNA ligase. The observed increase in the ratios of recombinant type and donor marker activity to resident marker activity is quantitatively due to an increase in the number of $trpC_2^+$ (donor marker) and $trpC_2^+ tyr_1^+$ (recombinant type) transformants after incubation with DNA ligase, the number of tyr_1^+ (resident marker) transformants in samples incubated with DNA ligase being equal to that obtained in the controls. The effect of incubation with DNA ligase on the recombinant type and donor marker activity is restricted to the DNA re-extracted very soon after the addition of calf thymus DNA; in some experiments it was found in the two minute sample only. To slow down the events following uptake, the transforming culture was shifted to 25°C after 10 minutes DNA uptake at 30°C. As can be seen from table I (columns I and II) donor marker and recombinant type activity re-isolated from the transforming complexes at the lower temperature are increased by DNA ligase to the same extent as observed at 30°C, but now the increase is observed for a substantially longer period of time. However, as soon as the donor marker activity has recovered from the eclipse phase (which coincides with the time that the maximal amount of recombinant type activity has been formed), no increasing effect of DNA ligase incubation upon the two types of activity is observed. Since single strand nicks strongly reduce the transforming activity of DNA, and repair of such nicks ef-

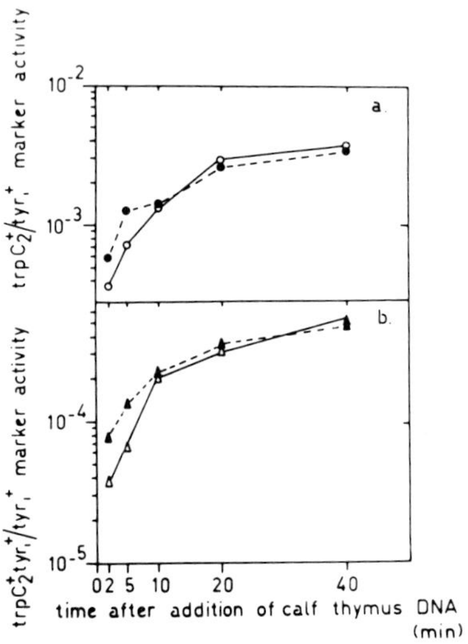

Fig. 1 - Effect of DNA ligase on (a) donor ($trpC_2^+$) and (b) recombinant ($trpC_2^+tyr_1^+$)type activity as a function of time of reisolation of the DNA.

Ratios of donor marker activity and recombinant type activity to resident marker activity were averaged from three independent experiments.

● -- ● and ▲-- ▲ : samples incubated with DNA ligase
0 — 0 and Δ — Δ : samples incubated with buffer.

fectively restores its biological activity (Laipis, Olivera and Ganesan, 1969), the effect of DNA ligase upon recombinant type and donor marker activity,observed during a limited and specific interval, strongly suggests the presence of single strand nicks in the donor-recipient complex.

TABLE I

EFFECT OF DNA LIGASE, DNA POLYMERASE AND EXONUCLEASE III ON THE RECOMBINANT TYPE AND DONOR MARKER TRANSFORMING ACTIVITY

Time after addition of calf thymus DNA (min)	$trpC_2^+/tyr_1^+$ (x 10^3)				$trpC_2^+$ tyr_1^+/tyr_1^+ (x 10^4)			
	I	II	III	IV	I	II	III	IV
2	2.2	3.4	3.7	3.3	2.0	4.4	4.4	4.7
5	3.7	5.2	5.2	5.3	3.6	6.7	6.9	6.4
10	4.5	6.0	5.8	6.8	5.4	9.1	9.5	9.0
30	6.7	9.4	9.0	9.4	12	18	19	19
50	9.3	9.2	9.9	9.0	19	20	19	17

I : controls (only buffer added).
II : + DNA ligase.
III : + DNA ligase + DNA polymerase
IV : + DNA ligase + DNA polymerase + exonuclease III.

To investigate whether in addition to recombining molecules with single strand nicks also recombining molecules with single strand gaps might be detectable in the re-extracted DNA, we also incubated reisolated DNA with DNA ligase+ DNA polymerase (table I,column III)and DNA ligase + DNA polymerase+ exonuclease III(table I,column IV).The reason for the addition of exonuclease III is that this enzyme is able to provide 3'OH endgroups required for polymerisation in the 5' to 3' direction of DNA polymerase both by virtue of its phosphatase and its exonuclease activity. As is shown in table I, no additional increase in donor marker and recombinant type is observed when the re-extracted DNA is incubated jointly with the three enzymes. The results either suggest that no recombining molecules with single strand gaps adjacent to the inserted segment of donor DNA are present in the re-extracted DNA, or in quantities too low to be detectable by the assay system used.

4. DISCUSSION

This paper shows that during the period in which integration proceeds recombinant type and donor marker activity can be increased by incubation with DNA ligase. With respect to the recombinant type activity it is clear that the DNA ligase is acting on an intermediate in the recombination process, because its effect on molecules carrying a donor and a resident marker is only observed for a limited, specific period of time. Slightly before (fig. 1) or as soon as (table I) the maximal number of biologically active recombinant molecules has been produced by the transforming culture, the effect of incubation of the re-extracted DNA with DNA ligase is no longer observed.

The increasing effect of incubation with DNA ligase on the re-extracted molecules carrying donor marker activity might be due to (1) repair of phosphodiester bond interruptions in non-integrated eclipse phase DNA, which carries a low amount of donor marker activity (Dubnau and Davidoff-Abelson, 1971) or (2) repair of such interruptions in recombining molecules in which both the linked donor $trpC_2^+$ and donor tyr_1 markers have not yet been covalently integrated. We favour the second explanation for the fol lowing reasons: (1) one would expect *a priori*, that recombining molecules carrying both the donor $trpC_2^+$ and the tyr_1(which will give rise to donor marker activity if integration is complete), will be similarly affected by DNA ligase activity as recombining molecules carrying only one (the $trpC_2^+$) donor marker. (2) There is a systematic difference in the relative increase of recombinant type activity and donor marker activity (2.0 and 1.4,respectively).This is precisely what would be expected if the DNA ligase is acting on a recombination intermediate endowed with future donor marker activity.In the increase of recombinant type activity subsequent to incubation with DNA ligase, one is observing the effect of sealing of a nick located in between the $trpC_2^+$ donor and the tyr_1^+ resident marker on joint transformation of that pair of markers, whereas in the increase of donor marker activity the effect of sealing nicks at any distance from the $trpC_2^+$ and tyr_1 pair of markers is observed. In the former case the nick is always close to the markers considered, whereas in the latter case the distance to the markers of a proportion of the nicks may be that large,that they do not interfere with their transforming activities. Consequently, sealing of such distant nicks will not enhance the transforming activity of the pair of donor markers.(3)The factorial increase in donor marker activity is approximately constant (1.4) throughout the period in which the increasing effect of DNA ligase is observed, which is difficult to reconcile with the possibility that the DNA ligase acts upon nicks in eclipse phase DNA,because the quantity of such DNA is steadily decreasing as a function of time of incubation of the transforming cells (Dubnau and Davidoff-Abelson, 1971).

After shifting the transforming cells to a lower temperature a delay is observed in the "cellular ligase" action. Dubnau and Davidoff-Abelson (1971) who have shifted cells from

37°C to 30°C, presumed the presence of a non-covalently bonded intermediate up to 50 minutes after stopping DNA uptake. This was based upon the specific transforming activity of ^3H labelled donor DNA. Venema et al. (1965 b) have reported that DNA with donor and recombinant type activity, reisolated soon after irreversible fixation, renatures less efficiently after heat denaturation than DNA obtained from cells in which recombination is completed. The present study, strongly suggesting the presence of a non-covalently bonded recombination intermediate in re-extracted DNA, together with the observation that the average size of integrated segments is relatively small with respect to resident DNA molecules (4.9×10^5, Bodmer, 1966, and 7.5×10^5, Dubnau and Davidoff-Abelson, 1971), may well explain the reduced capacity to renaturation. Effect of DNA ligase upon recombinant DNA structure has been described for T4 (Anraku and Lehman, 1969) and for λ DNA (Cassuto, Lash, Sriprakah and Radding, 1971). The intermediates in T4 recombination are incomplete joint molecules from which about 300-400 nucleotides are missing (Anraku, Anraku and Lehman, 1969). We have observed no additional increase in recombinant type and donor marker activity when the re-extracted DNA is incubated with DNA ligase, DNA polymerase and exonuclease III over that observed by DNA ligase only. There are two explanations for the absence of further increase of both types of activity upon additional incubation with DNA polymerase and exonuclease III. Firstly, in the recipient cells recombination intermediates with gaps adjacent to the inserted segments of donor DNA are present in quantities too low to be detectable by the methods employed. The second explanation is, that such intermediate molecules are not formed at all, because the assimilation of single stranded segments into the recipient chromosomes is effected by an exonuclease with properties similar to that of λ exonuclease described by Cassuto, Lash, Sriprakash and Radding (1971). We favour the latter possibility, because it is not immediately clear why the DNA-resynthesis step involved in the completion of gap carrying intermediates would be completed in a period of time much shorter than that required for linking the segment of donor DNA covalently to the resident moiety of DNA. However, in the absence of further data, no decisive conclusion can be drawn.

ACKNOWLEDGEMENTS

The authors thank Mr. H. L. Heijneker for DNA ligase, DNA polymerase and exonuclease III, Prof. W. J. Feenstra for a critical discussion and Mr. H.Beeftink and Miss Fien Helmers for expert technical assistance.The present investigations have been carried out under the auspices of the Netherlands Foundation for Chemical Research (S.O.N.) and with financial aid of the Netherlands Organization for the Advancement of Pure Research (Z.W.O.).

5. REFERENCES

ANRAKU, N. & LEHMAN, I. R. (1969). J. Mol. Biol. 46:467.

ANRAKU, N., ANRAKU, Y. & LEHMAN, I. R. (1969). J. Mol. Biol. 46:481.

BODMER, W. F. (1966). J. Gen. Physiol. 49:233.

BOTT, K. F. & WILSON, G. A. (1967). J. Bact. 94:562.

CASSUTO, E., LASH, T., SRIPRAKASH, K. S. & RADDING, C. M. (1971). Proc. Nat. Acad. Sci., Wash. 68:1639.

DUBNAU, D. & DAVIDOFF-ABELSON, R. (1971). J. Mol. Biol. 56:209.

FOX, M. S. (1966). J. Gen. Physiol. 49:183.

HEIJNEKER, H. L., PANNEKOEK, H., OOSTERBAAN, R. A., POUWELS, P. H., BRON, S., ARWERT, F. & VENEMA, G. (1971). Proc. Nat. Acad. Sci., Wash. 68:2967.

LAIPIS, P. J., OLIVERA, B. M. & GANESAN, A. T. (1969). Proc. Nat. Acad. Sci., Wash. 62:289.

NOTANI, N. & GOODGAL, S. H. (1966). J. Gen. Physiol. 49:197.

SPIZIZEN, J. (1958). Proc. Nat. Acad. Sci., Wash. 44:1072.

VENEMA, G., PRITCHARD, R. H. & VENEMA-SCHRÖDER, T. (1965 a). J. Bact. 89:1250.

VENEMA, G., PRITCHARD, R. H. & VENEMA-SCHRÖDER, T. (1965 b). J. Bact. 90:343.

RECOMBINATION

DNA SYNTHESIS INHIBITION BY 6-(p-HYDRO

IN RELATION TO UPTAKE AND INTEGRATION

IN BACILLUS SUBTI

BARBARA CHERNOV LEVI

OTTO E. LANDMAN

Department of Biology,
Georgetown University,
Washington, D. C. 20007, USA

ABSTRACT

The DNA synthesis inhibitor 6-(p-hydroxylphenylazo)-uracil (HPUra) was added to a competent culture of *Bacillus subtilis* 10 min before donor DNA and maintained in the transforming system until the bacteria were either plated for transformation(30 min)or lysed,preparatory to marker assay (90 min). Incorporation of ^{3}H thymidine into the recipients' DNA was inhibited 99.7% by HPUra. Uptake of radioactive donor DNA and formation of linkage between donor and recipient markers were both unaffected in the face of this inhibition of DNA synthesis. Transformant yield also was often unaffected by HPUra but was occasionally reduced to 50%. The hypothesis, that "an early step in transformation of *B. subtilis* is enhanced by chromosome replication", is held to be invalidated by these results. It is estimated, roughly, that the residual DNA synthesis which takes place in the presence of HPUra amounts to 2700 nucleotide pairs per chromosome. This synthesis is not discernibly affected by the presence of DNA and it can therefore not be wholly attributed to integration-related repair synthesis.

*Experiments performed by B.C.L. in partial fulfillment of the requirements of Georgetown University for the degree of Doctor of Philosophy.

1. INTRODUCTION

Erickson & Braun (1968) have shown that, in partially synchronized recipient cultures of Bacillus subtilis, the relative frequency of transformation for different markers may fluctuate markedly when DNA is added at successive times during the growth cycle. Moreover, the efficiency of transformation for a given marker was reported to be high around the time the replication fork was presumed to be passing its homolog in the recipient cells. These results were interpreted as indicating that integration of transforming DNA occurred preferentially at the replication fork (also see Bodmer, 1965) or, alternatively, that transport of a marker across the cell membrane was enhanced by the assistance of its homolog exposed at the fork.

More recently, Erickson & Copeland (1972) investigated the state of synchrony and the localization of the replication fork in the competent population fraction of recipient cultures. The rate of DNA synthesis in the competent fraction was lower and its chromosome alignment was different from that of the non-competent fraction. Nevertheless, the authors concluded that their results supported the hypothesis of Erickson & Braun "that an early step in the genetic transformation of B. subtilis is enhanced by chromosome replication".

Data analogous to those of Erickson have been recorded in abstracts from two other laboratories (Cooper & Evans, 1968; Sherratt & Collins, 1968).

The concept that transforming DNA in B. subtilis is integrated at the replication fork is not consistent with results reported by others: Laird, Wang & Bodmer (1968), studying physical integration of light donor DNA into ^{15}N ^{2}H labelled recipients growing in light medium showed by pycnographic analysis that integration occurred principally in preformed intermediate density (HL) regions of DNA and not in newly synthesized (LL) DNA. Ganesan (1967), in somewhat similar experiments, also reported that integration occurred randomly around the chromosome rather than in newly synthesized DNA.

The concept that the transport of a donor marker across the cell membrane is promoted by the presence of its resident homolog exposed at the fork was put to the test by Archer & Landman (1969 a & b). These workers studied transformation at three different loci around the chromosome in a recipient culture which had first been starved for amino acids to arrest the chromosomes at the terminus and then starved for thymine, in the presence of amino acids, for 8-10 hours to render it competent. In these synchronized, chromosome-arrested cultures the relative frequencies of transformation at the three loci were the same as in non-synchronized control cultures. The (presumed) absence of replication forks thus did not block marker uptake or force selective marker uptake. Moreover, protoplasting of the recipient culture after DNA uptake did not alter this picture qualitatively: apparently much of the donor DNA had passed into the protoplasts even while thymine starvation still maintained the recipients in chromosome arrest.

Collectively, the foregoing results argue rather forcefully against the idea that "an early step in the transformation of *B. subtilis* is enhanced by chromosome replication". Still, some uncertainties remain: concerning the experiments of Laird *et al.* and of Ganesan, one might posit that DNA transport was mediated by chromosome replication although integration was not replication dependent; concerning the experiments of Archer and Landman, it has been argued that, following unselective penetration of donor DNA past the cell wall in the chromosome-arrested recipient bacteria, the DNA might have been held sequestered there (within the protoplasts, or in the protoplast membrane) until chromosome replication resumed on the plating medium. Then, the postulated replication-dependent transport or integration step might have occurred (Erickson & Braun, 1968).

These issues might be resolved if one could study both uptake and integration under conditions where chromosome replication did not occur.

This approach was made possible by N. C. Brown's discovery that 6-(p-hydroxyphenylazo)-uracil (HPUra) completely blocks semi-conservative DNA synthesis in B.subtilis while still permitting repair synthesis (Brown, 1970; 1971). In this paper we show that both donor DNA uptake and integration (linkage formation) proceed unimpeded in the presence of HPUra concentrations which inhibit $[CH_3-^3H]$ thymidine incorporation into cellular DNA by 99.7%.

2. MATERIALS AND METHODS

(a) Overall experimental scheme

His^-trp^+ donor DNA is added to a competent his^+trp^- "first recipient" culture in presence or absence of the DNA synthesis inhibitor 6-(p-hydroxyphenylazo)-uracil (HPUra). After 20 or 30 min of incubation at 37°C, DNase is added and incubation continued for 5 min longer. Both recipient cultures are washed, protoplasted for 60 min in hypertonic medium and lysed. In the DNA-synthesis-inhibited culture these steps are carried out in presence of HPUra. The crude lysates are promptly assayed for donor (trp$^+$) and recombinant markers (his$^+$trp$^+$) on the his$^-$trp$^-$ "second recipient" strain. Cotransformation of hisB2 and trpC2 ranged between 66% and 80% in the present experiments.

The first recipient strain also carries the thyA$^-$thyB$^-$ markers and incorporation of $[CH_3-{}^3H]$-thymidine is measured along with transformation to monitor DNA synthesis. In other experiments, incorporation of ^3H labelled donor DNA into the first recipient is compared in the presence and absence of HPUra.

(b) Bacteria

All strains were derivatives of B. subtilis 168 (Spizizen, 1958). (Table I).

TABLE I

LIST OF STRAINS OF B. SUBTILIS

STRAIN NUMBER	MARKERS	CONTEXT OF USE	HOW OBTAINED
661	hisB2	donor DNA	Transformed # 540
665	hisB2 thyA thyB	Labelled donor DNA	Transformed # 664
664	trpC2 thyA thyB	First recipient	Farmer & Rothman via N. C. Brown
540	hisB2 trpC2	Second recipient	SB25, E. Nester

(c) Chemicals

HPUra, a gift from N.C. Brown, was stored at room temperature as a 20mM solution in 0.055M NaOH. Radioactive thymine (CH_3-^3H label, 11.4 Ci/mM)and thymidine(CH_3-^3H label 11.0 Ci/mM) were purchased from International Chemical and Nuclear Corporation. DNase and lysozyme were products of Worthington Biochemicals Corporation.

(d) Media

(i) RLS (Cahn & Fox, 1968) had the following composition: NH_4Cl, 0.08%; K_2HPO_4, 1.4%; K_2PO_4, 0.6%; sodium citrate, 0.1%; $MgCl_2.6H_2O$, 0.033%; Na_2SO_4, 0.0014%.

(ii) PLS was the same as RLS but also contained 0.5M sucrose. Sucrose was added on the day of use.

(iii) RGM was the same as RLS but also contained 0.5% glucose and the nine amino acids prescribed by Bott & Wilson (1967, 1968) at a concentration of 25 μg/ml each : L-histidine, L-arginine,L-valine, L-lysine,L-threonine, glycine,L-asparagine, L-methionine,L-tryptophan.Amino acids required by the auxotrophs as supplements were added at a concentration of 50 μg/ml to the first growth medium and at a concentration of 25 μg/ml to the second growth medium. Thymine was added at a concentration of 20 μg/ml, except where noted.

(iv) PGM was the same as RGM but also contained 0.5M sucrose.

(v) Minimal agar(Spizizen, 1958)had the following composition: K_2HPO_4, 1.4%; KH_2PO_4, 0.6%; sodium citrate, 0.1%; $(NH_4)_2SO_4$, 0.2%; $MgSO_4 \cdot 7H_2O$, 0.02%; glucose, 0.5%; agar (Difco), 1.5%; amino acids, when required, 0.005%; thymine, when required, 0.002%.

(vi) Dilutions were made in RLS or, when osmotic stabilization was required, in DFL. DFL is the same as PLS but also contains 0.01M $MgCl_2$.

All media was prepared with distilled, deionized water.

(e) Competence procedure

Cells were grown overnight in Penassay Broth (Difco Antibiotic Medium no. 3) at 37°C with shaking. Thymine requiring strains were supplemented with 20 µg/ml thymine. The cells were centrifuged and resuspended in freshly prepared RGM to a Klett reading of approximately 16 (Klett-Summerson colorimeter, filter #42). The culture was placed on a rotary shaker at 37°C and the turbidity monitored at hourly intervals.

Two hours after reaching the end of log phase, the cells were centrifuged, resuspended to 1/4 volume in PGM and subdivided into 1 ml samples. These were quickly frozen in an acetone-dry ice bath, then stored at -20° C. Frozen cells were used within two weeks of preparation. Loss of viability and competence were negligible during this time. For use in an experiment, the cells were thawed rapidly, diluted 1:3 with PGM or RGM and allowed to grow at 37° C with shaking until they reached a turbidity which had previously been determined as corresponding to the peak of competence. This required about two hours for strain 540 and about 3 1/2 hours for strain 664. $MgCl_2$ at a final concentration of 0.02M and then DNA were added. In first recipient transformations, DNA was added at a concentration of 5μg/ml,incubation was allowed to proceed at 37°C with shaking for 20min and transformation was then terminated by addition of DNase(10 μg/ml final concentration). In second recipient transformations, DNA--in crude cell lysates--was added at several dilutions to insure quantitative assay.Transformation of the second recipient was allowed to proceed at 37°C for 30min.The recipient culture was concentrated five-fold by centrifugation-resuspension prior to plating to obtain more transformed colonies per plate. All platings were done in triplicate.

(f) DNA extraction

DNA from donor cultures was extracted by the method of Marmur (1961). DNA concentration was determined by the diphenylamine reaction (Burton, 1956).

The DNA of the first recipients was assayed in crude lysates which were prepared as follows:following the transformation incubation, with or without HPUra,transformation was terminated by addition of 10 μg/ml DNase. The cells were centrifuged at 9,000 x g for 10 min at 4°C and washed twice. They were then suspended in DFL containing 200 μg/ml lysozyme and 160 μM HPUra (HPUra diluent in the controls). Protoplasting was allowed to proceed for 60 min in a 37°C water bath. Both microscopic examination and plating of samples were used to check for completeness of protoplasting. The protoplasts were centrifuged at 2,000 x g for 10 min at 4° C. A slurry was formed by stirring 0.04 ml DFL (±HPUra) into the protoplast pellet. The protoplasts were then lysed by the rapid addition of 2 ml cold sterile distilled water.

The crude lysates were kept on ice no longer than 30 min before use in the transformation of the second recipient.

(g) Preparation of radioactive DNA

In the course of this work, it was discovered that, under some conditions, namely the presence of 20 μg/ml cold thymine, exogenous DNA markedly enhanced the uptake of labelled thymidine into the alkali-stable, cold trichloroacetic acid (TCA) precipitable fraction of strain 664. Homologous DNA was somewhat more effective than calf thymus DNA. Deoxyadenosine did not stimulate thymidine uptake. Radioactive DNA was prepared as follows: A frozen, concentrated, pre-competent culture of strain 665(his^-thy^-) was thawed, diluted and grown with 1 μg/ml thymine until the peak of competence was reached as previously determined. 5 μg/ml homologous thy^-his^- DNA and 0.1 μC(CH_3-^3H) thymidine/ml were added to the culture and growth continued at 37°C with shaking. Another 0.1 μC(CH_3-^3H)thymidine/ml was added after 30 min. The cells were harvested after 30min of additional growth and DNA was extracted according to Marmur's procedure (1961). Final radioactivity was 2×10^4 cpm/μg DNA.

(h) Incorporation of labelled thymidine

Incorporation of (CH_3-^3H)thymidine into DNA was determined by radioactivity assay of the alkali-stable, cold TCA insoluble portion of whole cells(Roodyn & Mandel, 1960). Triplicate samples from single or duplicate flasks were impinged on Whatman GF/C filters presoaked in 5% TCA containing 200 μg/ml thymidine (filters were dried before use). Then repeated washes were performed with cold 5% TCA containing thymidine,cold 1% TCA containing thymidine, and cold 1% acetic acid.Finally, the filters were rinsed in cold 95% ethanol to hasten drying.

In a few experiments, the alkali stable, cold-TCA-extracted pellets were subjected to further fractionation : they were held in 5% TCA at 90° C for 30 min to hydrolyze DNA. Some DNA-extracted samples were washed with 0.01 M NH_4HCO_3, pH8, and treated in this buffer with 1 mg/ml trypsin for 120 min at 37°C to remove protein. Some trypsin-treated samples were washed and incubated in 0.01 M NH_4HCO_3 with 1 mg/ml lysozyme for 120 min at 37° C to remove cell wall (the foregoing procedure was modified after Brown & Handschumacher 1966). In each case, the insoluble residues remaining after treatment were carefully collected on GF/C filters and washed extensively. The filters were then dried and counted in 5 ml of scintillation fluid containing 0.5% 2,5 diphenyloxazole and 0.02% 1,4-bis-2-(5-phenyloxazolyl)-benzene in toluene. Counting was in a Packard case on try-carb liquid scintillation counter, model 3310.

3. RESULTS

(a) Effect of HPUra on (CH_3-^3H) thymidine uptake by competent cultures in presence and absence of DNA

Although N.C.Brown had shown earlier that net DNA synthesis and uptake of labelled thymidine for DNA replication are both negligible in the presence of 80 µM HPUra (Brown, 1970), it was necessary to show that DNA synthesis inhibition was also effective in competent cultures and in the presence of transforming DNA.

Parallel cultures of competent cells were incubated for 10 min with shaking in the presence or absence of 160 μM HPUra. ^{3}H thymidine was then added, and 5μg/ml donor DNA where indicated, and incubation continued 30 min longer. The cultures were then plated and sampled for radioactivity. Suspensions of heat killed cells (75°C 30min) were incubated, washed and extracted in parallel with the experimental samples to serve as controls. The results show (Table II) that HPUra inhibited ^{3}H thymidine incorporation with about 99.7% effectiveness while affecting transformation efficiency only slightly. Inhibition is similar when thymine is used as the tracer. In the uninhibited samples, DNA often enhanced thymidine incorporation slightly. This stimulation is also produced by DNase digests of DNA and by calf thymus DNA and is therefore probably attributable to DNA breakdown products (Firshein, 1965).

(b) Effect of HPUra on uptake of labelled DNA

Competent cells were incubated with 160 μM HPUra or diluent for 10 min. ^{3}H labelled DNA (5 μg/ml) was added and incubation continued for 30min at 37°C with shaking. Transformation was then terminated by the addition of DNase (10 μg/ml, 5 min incubation at 37°C with shaking). The cultures were centrifuged, washed twice to remove exogenous radioactivity (9000 x g, 10 min at 4°C) and then sampled for radioactivity assay and transformation (plating) assay. HPUra was maintained at 160 μM in the "experimental" samples throughout all washings. The results of the initial experiment are shown in row A of Table III. Although the transformation frequency was the same in the HPUra sample as in the control sample the apparent uptake of radioactivity was about 30% lower. Since recipients are known to break down DNA and use

TABLE II

EFFECT OF HPUra ON $[CH_3-{}^3H]$ THYMIDINE UPTAKE AND ON TRANSFORMATION

EXPERIMENT #	HPUra 160 μM	DNA 5 μg/ml	% THY^+ TRANSFORMANTS[f)]	cpm/0.2 ml[a)]	HOT TCA DEGRADABLE cpm/0.2 ml	% INCORPORATION OF CONTROL
G	-	-	-	11,176[b)]		100
	+	-	-	30[c)]		0.27
	-	+	.618	13,101[b)]		100
	+	+	.357	35[c)]		0.27
H	-	-	-	37,033[b)]		100
	+	-	-	248[b)]		0.67
	-	+	.532	43,485[b)]		100
	+	+	.261	226[b)]		0.52
I	-	+	.562	82,730[c)]	80,559[d)]	100
	+	+	.311	494[c)]	318[d)]	0.40

J[e]	+	-	-	1,938[c]	1,373[d]	
	-	+	.258	198,243[c]	193,887[d]	100
	+	+	.219	979[c]	479[d]	0.25

a) Corrected by activity of heat killed cell controls. Not corrected for tracer incorporated into protein.

b) Average of 3 samples from one tube.

c) Average of two tubes, each sampled 3 times.

d) Average of two tubes, each sampled 2 times. All radioactivity was in DNA.

e) Sampled for radioactivity 90 min after DNA addition instead of 30 min.

f) In experiments A-E (Tables III and IV) the cells were grown on 20 μg/ml thymine in both growth media. In experiments G-J, the second growth medium contained only 3 μg/ml thymine in order to minimize isotope dilution. It is not known whether this difference accounts for the moderate inhibition of transformation by HPUra in the latter experiments. Transformation frequencies for the $\underline{trp}^+$ marker were also lower in the HPUra inhibited samples of experiments G, H and I.

TABLE III

EFFECT OF HPUra ON DNA UPTAKE AND TRANSFORMATION[a)]

EXPT #[b)]		HPUra	50 μg/ml THYMIDINE	cpm/ml	% TRP^+ TRANSFORMANTS	% TRP^+ HIS^+ TRANSFORMANTS
A		-	-	1585	.32	.10
		+	-	1145	.32	.11
B	1	-	-	1130	.29	.10
		+	-	905	.29	.10
	2	-	+	980	.24	.08
		+	+	960	.28	.09
C		-	+	535	.99	.27
		+	+	615	.62	.21

a)Radioactivity measurements were made on the cold-TCA-insoluble, 0.1 N alkali insoluble fraction of the recipient population (Roodyn & Mandel,1960).The counts shown are averages from triplicate samples. In experiment 3, a sample of heat killed cells was included as a control--its radioactivity was at background level.

b)The same batch of frozen cells was used in experiments A and B.

the breakdown products for new DNA synthesis (Lacks, Greenberg & Carlson, 1967) 50 μg/ml cold thymidine was added in the next experiment (B) to see whether the difference between the DNA-synthesizing and the HPUra-containing suspensions could be eliminated. Rows B and C of Table III show that in the presence of cold thymidine radioactive DNA uptake was the same in the presence as in the absence of HPUra.

(c) Effect of HPUra on linkage formation

The preceding experiments show that donor DNA uptake occurs with normal efficiency under circumstances when DNA synthesis is almost completely blocked. But what about DNA integration? Does it proceed in DNA-replication-inhibited cells or does it occur only after dilution and plating?To answer this question we compared linkage formation in HPUra-inhibited strains with linkage formation in DNA-replicating control cultures.

Competent thymine requiring <u>his</u>$^{+}$ <u>trp</u>$^{-}$ cultures were incubated 10 min at 37° C with shaking in presence or absence of 160 μM HPUra.Then 5 μg/ml his^{-} trp^{-} DNA was added and incubation continued for 20 min. DNA uptake was terminated by 5 min of incubation with 10 μg/ml DNase and samples plated to assay for transformation of the "first recipient". ^{3}H thymidine uptake was followed alongside in separate flasks to monitor DNA synthesis and its inhibition by HPUra. The main cultures were washed twice to remove DNase,resuspended in DFL containing 200 μg/ml lysozyme and incubated 60min at 37°C.The resultant protoplast suspensions were centrifuged and the pellets lysed by the addition of 2 ml cold water. In the HPUra samples the inhibitor was maintained at 160 μM throughout the incubations with DNA, lysozyme and the centrifugations up to the time of lysis (about 90 min).Dilutions of the crude lysates of the HPUra samples and the control samples were assayed promptly on the <u>trp</u>$^{-}$<u>his</u>$^{-}$ strain for the donor marker <u>trp</u>$^{+}$ and for linkage formation with first recipient(<u>trp</u>$^{+}$<u>his</u>$^{+}$). Results of three experiments are shown in Tables IVA and IVB . Table IVB shows that, in each of the experiments, linkage formation was similar in the HPUra samples and the control samples (columns 6, 7 and 8)(in a fourth experiment,2nd recipient transformations were erratic but were not indicative of any inhibition by HPUra).

TABLE IVA

EFFECT OF HPUra ON LINKAGE FORMATION: TRANSFORMATION AND DNA UPTAKE IN THE FIRST RECIPIENT

EXPERIMENT #	HPUra	% TRP^+ TRANSFORMANTS	$\frac{TRP^+HIS^+}{TRP^+}$	% COMPETENT CELLS[a)]	^3H THYMIDINE INCORPORATION cpm/0.2 ml[b)]
D	-	0.28	0.30	-	5427
	+	0.28	0.27	-	117
E	-	0.32	0.28	11.7	3438
	+	0.30	0.27	10.9	28
F	-	0.56	0.21	25.1	4040
	+	0.47	0.20	20.7	34
COLUMN 1	2	3	4	5	6

a) Calculated on the basis of single $\underline{trp}^+$ and $\underline{thy}^+$ transformants and $\underline{trp}^+\underline{thy}^+$ doubles.

b) Data are corrected by subtraction of heat-killed cell activity;they are not corrected for counts incorporated into protein --which, in the case of HPUra inhibited samples,account for about half of the radioactivity. Counts are averages of data from 1 or 2 tubes, each sampled in triplicate.

How far had integration proceeded in the first recipients 90 min after DNA addition, at the time they were lysed? An approximate answer to this question can be obtained by comparing linkage formation after 90 min with ultimate linkage formation in 1st recipient transformations. In both cases the ratio integrated $\underline{trp}^+$ markers(i.e. $\underline{trp}^+\underline{his}^+$ doubles) to total $\underline{trp}^+$ markers serves as the measure of integration. For the 90 min time point, this ratio is obtained from the assay of the lysate on the second

TABLE IVB

EFFECT OF HPUra ON LINKAGE FORMATION: TRANSFORMATION ASSAYS OF LINKAGE WITH THE SECOND RECIPIENT

EXPERI-MENT #	HPURA	% TRANSFORMATION						$\frac{TRP^+ HIS^+}{TRP^+}$			ESTIMATED COMPLETION OF LINKAGE %
		TRP^+ $x10^4$			HIS^+TRP^+ $x10^5$						
		DILUTIONS			DILUTIONS			AT DILUTIONS			
		10^0	10^1	10^2	10^0	10^1	10^2	10^0	10^1	10^2	
D	-	.49	.21	-	.62	.19	-	.13	.09	-	62[a)]
	+	.49	.07	-	.60	-	-	.12	-	-	61
E	-	-	3.0	-	-	6.2	-	-	.20	-	100
	+	-	4.9	-	-	11.7	-	-	.24	-	122
F	-	-	8.9	.81	-	9.6	.71	-	.11	.09	59
	+	-	10.7	.80	-	11.2	1.07	-	.10	.13	63
COLUMN 1	2	3	4	5	6	7	8	9	10	11	12

a)This figure was calculated as follows: cotransformation of <u>his</u> and <u>trp</u> was 70% in the donor--first recipient transformation, resulting in 30% $\underline{his}^+\underline{trp}^+$ transformants (Table IVA, column 4). If cotransformation is assumed to be the same in the transformation of the second recipient, the expected percentage of $\underline{his}^+$ $\underline{trp}^+$ transformants among all $\underline{trp}^+$ transformants is 21%. The actual percentage (Table IVB, column 9) was 13%, i.e. 62% of expected.

recipient (Table IVB, columns 9, 10 and 11). For "ultimate linkage" estimations, the first recipient transformation data are used (Table IVA, column 4).The comparison of these ratios (Table IVB,column 12) suggests that integration at 90 min is rather far advanced.The high ratios of $\underline{his}^+\underline{trp}^+$ to total $\underline{trp}^+$ in the 90 min lysates indicate that most $\underline{trp}^+$ donor DNA destined for destruction is no longer active as transforming DNA at that time. HPUra did not impede integration in any of these experiments.

(d) Repair synthesis in the transformation of B. subtilis

Although the data of Tables II, III, IVA and IVB show that 99.7% inhibition of bacterial chromosome replication does not materially affect DNA uptake or integration, no clear conclusions were drawn concerning the involvement of repair synthesis in transformation. To obtain some insight into this question we examined more closely the residual 0.3-0.4% uptake of label in HPUra-inhibited cultures. In experiments I and J (Table II) the alkali-stable, cold-TCA-precipitated samples were extracted successively with hot TCA, trypsin and lysozyme to ascertain whether the radioactivity was indeed in DNA. As shown in Table II, a substantial part of the radioactivity in the HPUra-treated cells was contained in DNA since it could be extracted by hot TCA (virtually all of the remaining radioactivity was solubilized by trypsin). Are these residual counts due to leakage through the HPUra block--do they represent general repair synthesis or repair synthesis related to DNA integration ? A comparison of residual incorporation in presence and absence of donor DNA in 14 experiments suggests that much--and perhaps all--of this radioactivity is due to incomplete inhibition by HPUra and not integration-related. In particular, in 7 out of these 14 experiments there was less or the same radioactivity in the samples with donor DNA than in the samples without added DNA.

A more reliable assessment of the contribution of repair synthesis to the residual radioactivity of HPUra-inhibited cells was attempted in the following experiment. After allowing transformation and ^{3}H thymidine incorporation in the presence of HPUra, a culture was fractionated with a DEAE column into competent(10%)and noncompetent (90%) subpopulations(for a description of this method see Archer & Landman, 1969b)--then the residual radioactivity of the two population fractions was measured. Very preliminary results indicate that the radioactivity per cell is the same or lower in the competent subpopulation than in the non competent cells.

Viable counts, OD measurements and DNA analyses indicate that the DNA content of our transforming cultures increases no more than 20% during the 90 min experimental period between donor DNA uptake and lysis preparatory to DNA assay. Thus, for example, the 193,887 counts accumulated during the 90min of experiment J (Table II) represent at most a 20% increment in chromosomal DNA. The 479 counts accumulated under HPUra inhibition accordingly represent 0.0005 of chromosomes synthesized or about 2700 nucleotide pairs per chromosome(based on Erickson's estimate of a chromosome size for B. subtilis of 3.3 x 10^9 Dalton--see Erickson, 1970).

If our preliminary (DEAE column) observations are confirmed that transforming cells incorporated no more radioactivity under HPUra inhibition than noncompetent cells, an important proportion of these 2700 nucleotide pairs must be attributable to leakage through the HPUra block rather than to integration-related repair synthesis.

4. DISCUSSION

One of the important objectives of the present research was to test the proposition of Erickson and Copeland (1972) that "an early step in the genetic transformation of B. subtilis is enhanced by chromosome replication". Our experiments focussed on the two "early" steps in transformation which are plausibly connected to the recipient's chromosome replications, namely donor DNA uptake and donor DNA integration. Our results show that in the face of 99.7% inhibition of DNA synthesis by HPUra, both uptake (Table III) and linkage formation (Table IVB) took place with normal efficiency.

Assuming ^{3}H thymidine incorporation occurred uniformly throughout the recipient population, we may visualize the replication fork sweeping past 20% of the chromosome in the control culture and creeping past 0.05% of the chromosome - 2 or 3 genes - in the HPUra culture during the 90 minute experimental period (assuming that the residual incorporation represented semiconservative synthesis). If exposure of a marker at the fork enhanced either the transport of its homolog through the membrane or its integration at the fork, surely the vastly more frequent exposure of this marker in the uninhibited culture would be reflected in markedly increased uptake or integration!

In the abstract, one could object to the foregoing picture with the argument that the competent fraction of the recipient population might have incorporated more than its proportionate share of the residual radioactivity, i.e. that it was synthesizing more DNA per cell than the noncompetent population fraction. In actuality, the experiments of Nester's group have shown quite convincingly that the competent population fraction is much less active in DNA synthesis than the noncompetent fraction (McCarthy and Nester, 1967). They demonstrated in "^{3}H suicide" experiments that, immediately after DNA uptake, 50% of the transformed cells do not take up measurable (i.e. suicide-producing) amounts of ^{3}H thymidine and that the remaining 50% of the transformants take up much less ^3H thymidine than the noncompetent population fraction. These results have since been confirmed and extended in several laboratories (Dooley, Hadden and Nester,1971, Erickson & Copeland, 1972).

Preliminary experiments in our HPUra system on transformed and noncompetent cell fractions separated on a DEAE column also indicate the DNA synthesis in the competent population is not accelerated.

In rejecting the hypothesis that bacterial chromosome replication enhances DNA uptake or DNA integration,we are led to face the question why transformation frequencies for different markers fluctuate in partially synchronized recipient cultures (Erickson and Braun, 1968). Since our investigation of DNA synthesis inhibition up to the time of donor DNA integration produced no hint of an explanation, our attention is directed to events following integration. Traditionally, it has been widely assumed that integration of donor DNA is the terminal event in transformation.However,there is substantial evidence that transformation-related events occur in recipient cells 2-5 hours after donor DNA uptake. (Nester & Stocker, 1963, Kammen, Wojnar & Canellakis,1966,Vermeulen & Venema,1972,Knott & Landman,unpublished). Particularly clear-cut is the finding by Davidoff-Abelson and Dubnau (1971) that, in Rec B2 mutants of B. subtilis, linkage formation between trp^+ donor DNA and his^+ recipient DNA occurred normally but that,subsequent to integration,the trp^+recombinants failed to multiply. Finally, in an analysis of genetic heterogeneity among transformant clones,Bresler,Kreneva & Kushev(1968) found that the frequency of pure transformant clones among pure and mixed clones varied for the six different markers used between 18 and 80 percent. This finding strongly suggests site-specific strand correction in heteroduplex regions of B.subtilis, analogous to the mechanism operating in low efficiency transformation in pneumococcus (Ephrussi-Taylor & Gray, 1966). We believe such a correction mechanism could explain the results of Erickson & Braun (1968) , provided that correction can be shown to be influenced by the state of synchrony or the metabolic state of the cell. (We have learned from Dr. J. F. Collins that the fluctuations in transformability of different markers observed by Sherratt & Collins (1969) are largely suppressed by adenine supplementation).

One of the most useful findings reported in this paper is that marker recombination in transformation can be studied in a fully viable system exhibiting a very low background level of DNA synthesis. We estimate the residual incorporation amounts to roughly 2700 nucleotide pairs per chromosome. It is not yet possible to judge how much this residual incorporation is due to leakage through the HPUra block (Bazill & Gross, 1972) and how much to repair processes associated with marker integration.

ACKNOWLEDGEMENTS

We thank Dr. Neil C. Brown (Baltimore) for helpful discussions and for his gifts of HPUra and of a bacterial strain.

This investigation was supported by grant # AI 05972 from the National Institute of Allergy and Infectious Diseases and grant GB 27607 from the National Science Foundation to O.E.L. B. C. L. was supported by Environmental Health Training grant ES 00087 from the National Institute of Environmental Health Sciences.

5. REFERENCES

ARCHER, L. & LANDMAN, O. E. (1969a). J. Bact. 97:166.

ARCHER, L. & LANDMAN, O. E. (1969b). J. Bact. 97:174.

BAZILL, G. W. & GROSS, J. D. (1972). Nature New Biology, 240:82.

BODMER, W. (1965). J. Mol. Biol. 14:534.

BOTT, K. F. & WILSON, G. A. (1967). J. Bact. 94:562.

BOTT, K. F. & WILSON, G. A. (1968). Bact. Rev. 32:370.

BRESLER, S. W., KRENEVA, F. A. & KUSHEV, V. V. (1968). Molec. Gen. Genetics, 102:257.

BROWN, N. C. (1970). Proc. Nat. Acad. Sci., Wash. 102:257

BROWN, N. C. (1971). J. Mol. Biol. 59:1.

BROWN, N. C. & HANDSCHUMACHER, R. E. (1966). J. Biol. Chem. 241:3038.

BURTON, K. (1956). Biochem. J. 62:315.

CAHN, F. H. & FOX, M. S. (1968). J. Bact. 95:867.

COOPER, W. G. & EVANS, J. E. (1968). Biophys. J. Abstr. 8:A-79.

DAVIDOFF-ABELSON, R. & DUBNAU, D. (1971). Proc. Nat. Acad. Sci. 68:1070.

DOOLEY, D. C., HADDEN, C. T. & NESTER, E. W. (1971). J. Bact. 108:668.

EPHRUSSI-TAYLOR, H. E. & GRAY, T. C. (1966). J. Gen. Physiol. 49:part 2, 211.

ERICKSON, R. J. (1970). Curr. Topics Microbiol. Immunol. 53:149.

ERICKSON, R. J. & BRAUN, W. (1968). Bact. Rev. 32:291.

ERICKSON, R. J. & COPELAND, J. C. (1972). J. Bact. 109:1075.

FARMER, J. L. & ROTHMAN, F. (1965). J. Bact. 89:262.

FIRSHEIN, W. (1965). J. Bact. 90:327.

GANESAN, A. T. (1967). In Organizational Biosynthesis, ed. by H. J. Vogel, J. O. Lampen & V. Bryson. New York and London: Academic Press.

KAMMEN, H. O., WOJNAR, R. J. & CANELLAKIS, E. S. (1966). Biochim. biophys. acta. 123:56.

LACKS, S., GREENBERG, B. & CARLSON, K. (1967). J. Mol. Biol. 29:327.

LAIRD, C. D., WANG, L. & BODMER, W. F. (1968). Mutation Res. 6:205.

MARMUR, J. (1961). J. Mol. Biol. 3:208.

McCARTHY, C. & NESTER, E. W. (1967). J. Bact. 94:131.

NESTER, E. W. & STOCKER, B. A. D. (1963). J. Bact. 86:785.

ROODYN, D. B. & MANDEL, H. G. (1960). Biochim. biophys. acta. 41:80.

SHERRATT, D. J. & COLLINS, J. F. (1969). J. Gen. Microbiol. 55:x.

SPIZIZEN, J. (1958). Proc. Nat. Acad. Sci., Wash. 44:1072.

VERMEULEN, C. A. & VENEMA, G. (1972). J. Gen. Microbiol. 71:415.

ISOLATION AND CHARACTERIZATION OF RECOMBINATION DEFECTIVE MUTANTS OF BACILLUS SUBTILIS

NIGEL HARFORD

IGOR SAMOJLENKO *

MAX MERGEAY

Cell Biochemistry, Radiobiology Department
CEN/SCK, B 2400 MOL, Belgium

ABSTRACT

A number of radiation sensitive mutants were isolated after mutagenesis of Bacillus subtilis. Three of these mutants (recB19, rec-27, rec-61) are recombination defective as measured by the uptake and expression of ^{3}H-labelled DNA relative to wild type. The recB19 mutant retains about 5% and mutants rec-27 and rec-61 about 10-20% of the wild type level of transformability as judged by this test. In common with other rec^{-} strains the mutants display increased sensitivity to u.v. and X-ray irradiation but are able to effect HCR of u.v.-irradiated phage. Breakdown of DNA after u.v.-irradiation is increased in rec-27 and rec-61 compared to wild type but not in recB19.

The extent of repair of u.v.-induced damage to prelabelled DNA was estimated by centrifugation on alkaline sucrose gradients. Some capacity for repair of u.v.-lesions in DNA is retained by the mutants recB19 and rec-61 while rec-27 and a mutant tentatively classified as EXR (exr-15) show little or no repair ability.

* Permanent address: Gamalaya Institute of Microbiology,
MOSCOW D-98
U.S.S.R.

1. INTRODUCTION

A number of mutants of Bacillus subtilis defective in transformation and/or recombination ability have been isolated in several laboratories (Searashi and Strauss, 1965; Okubo and Romig, 1966; Hoch, Barat and Anagnostopoulos, 1967; Hoch and Anagnostopoulos, 1970; Zadrazil and Fucik, 1971; Hill, Prakash and Strauss, 1972; and see Polsinelli; Yasbin, Tevethia, Wilson and Young; these proceedings). Defective strains can be subdivided into two main categories. Firstly those blocked at some stage relating to the entry or uptake of DNA into the cell, or in the development of competence; secondly those where DNA uptake is normal or proficient but which are defective in some step leading to the integration (or expression) of donor information. Most of the mutants in this second class which have been isolated so far are pleiotropic with respect to their phenotypic properties, as are similar mutants isolated in other transformable or conjugable bacterial species (Clark and Margulies, 1965; Holloway, 1966; Postel and Goodgal, 1972; Kooistra and Venema, 1970; Beattie and Setlow, 1971). In addition to being poorly transformable or conjugable they are sensitive to u.v.-and/or X-irradiation, to radiomimetic chemicals such as MC and MMS* and may also show reduced levels of transfection or transduction as well as other properties. These mutants are of interest as they may provide information on the enzymatic mechanisms and steps in

* Abbrevations used: MC, mitomycin C; MMS, methylmethane sulphonate.

the pathway(s) leading to integration of incoming donor DNA, repair of genetic damage, maintenance of genome integrity and the interrelationships between these processes. Transformable bacterial species offer certain advantages for the investigation of recombinational processes as the molecular fate of donor DNA can be directly studied by both physico-chemical and genetic techniques.

We report here on some properties of five X-ray and/or u.v.-sensitive mutants of *B. subtilis* and on their ability to repair single strand break damage to DNA after irradiation with u.v.-light. Three of these mutants (*rec*B19, *rec*-27, *rec*-61) appear to fulfill the criteria of recombination defectiveness while a fourth mutant (*exr*-15) has the characteristics of an EXR or LEX phenotype (Rorsch *et al.*,1966; Howard-Flanders and Boyce, 1966). The fifth mutant (*uvr*-56) appears defective in some step in photodimer excision and repair.

2. MATERIALS AND METHODS

(a) Bacterial strains

The mutant strains of *B.subtilis* are listed in Table I. Strains GSY228, GSY1027 and GSY1028 were obtained from the collection of Dr. C. Anagnostopoulos. GSY1027 and GSY1028 carry respectively the *uvr*-1 and *rec*B2 mutations originally described by Reiter and Strauss (1965) and Okubo and Romig (1966). Mutations in the other strains listed were isolated in this laboratory by nitrosoguanidine mutagenesis of strain GSY228 and selection for sensitivity to MC or MMS. Mutants were purified by cloning and reintroduced into GSY228 by transformation.

(b) Ultraviolet irradiation

Ultraviolet survival curves were obtained by dilution of overnight Penassay broth cultures in a Tris-mineral salts buffer and exposure to a 6W germicidal lamp at a distance of 60 cm. Survival was estimated by plating on Tryptose Blood Agar Base medium (TBAB).

(c) DNA breakdown following u.v.-irradiation

Cultures were grown overnight in Penassay both supplemented with 250 μg/ml. deoxyadenosine and 20-30 μc [^{3}H] thymidine/ml. (CEN/SCK, Mol; spec. act. 10-12 Ci/m-mole). After washing the cells were diluted in CT5 medium (minimal salts medium of Anagnostopoulos and Spizizen (1961) supplemented with 0,05% tryptose and 50 g/ml. required amino acids) and grown to an optical density at 660 mμ of 0.3 or about 1 x 10^{8} cells/ml. for GSY228. Half the culture (8 ml.) was exposed to u.v.-light for 150 seconds before reincubation at 37°. Samples from control and irradiated cells were precipitated with 5% TCA, centrifuged and 0.2 ml. aliquots of the supernatant counted to determine the release of acid soluble material.

(d) Transformation and uptake of ^{3}H-labelled DNA

DNA labelled with [^{3}H] thymidine was isolated from a thy derivative of B. subtilis 168 (Farmer and Rothman, 1965) by pronase digestion of cell lysates in 1% sodium dodecylsulphate followed by phenol extraction, alcohol precipitation and solution in SSC. The specific activity of these DNA preparations was generally about 1 - 3 x 10^{6} d.p.m./ μg.

Cultures were grown for competence by the two phase regimen using tryptose as the growth supplement (Harford and Sueoka, 1970). Uptake and transformation were tested by incubating 1 ml. competent culture with 1 - 2 μg ^{3}H-DNA for 15 minutes at 37°. The reaction was stopped by addition of 6 ml. ice cold medium containing 10 μg/ml. deoxyribonuclease and chilling. Cells were washed three times by centrifugation and samples taken for estimation of *trp*$^{+}$ transformants and uptake of radioactivity.

(e) Alkaline sucrose gradient centrifugation

Cultures were labelled with 50 μc/ml [^{3}H] thymidine and u.v.-irradiated as in (c) above. Samples of 1 ml. were taken immediately before and after irradiation and after 30 and 100 minutes post-irradiation incubation.

The samples were diluted into cold Tris-NaCl-EDTA-NaN_3 buffer (pH 8.0), centrifuged and resuspended in the same buffer with 500 μg/ml. lysozyme. Partial protoplasts were formed by incubation for 3-4 minutes at 37° and 0.1 ml. layered gently on 0.1 ml. 0.5 *M*-NaOH on top of a 5-20% alkaline sucrose gradient in 0.9 *M*-NaCl, 0,1 *M*-NaOH resting on a 0.5 ml. cushion of 60% sucrose.

Gradients were centrifuged at 30,000 rev./min. at 20° for 1.5 hours in a Martin Christ 60 rotor. Fractions were collected by dripping onto Whatman GF/A glass fibre filters and washing these successively twice with 5% TCA and once in ethanol and acetone. Recovery of the total radioactivity from the gradients was between 75 to 95%.

TABLE I

GENERAL PROPERTIES OF RECOMBINATION DEFECTIVE AND OTHER MUTANTS OF B. SUBTILIS

Strain designation	Genotype	Sensitivity [a] to u.v.	Sensitivity [a] to X-ray	HCR Capacity [b] $\frac{Nuv}{No}$	Transformability relative to GSY228 (per cent)	Retains some capacity to repair u.v. damage to DNA [c]
GSY228	trpC2, metB4	+	+	6.7×10^{-2}	100	yes
1619	trpC2, metB4, recB19	s	s	8.4×10^{-2}	5.4	yes
1627	trpC2, metB4, rec-27	s	s	5.0×10^{-2}	19.4	no
1661	trpC2, metB4, rec-61	s	s	7.0×10^{-2}	18.5	yes
1615 1618	trpC2, metB4, exr-15 trpC2, metB4, exr-18	s	s	8.4×10^{-2}	39	no
1027	trpC2, metB4, uvr-1	s	+	2.6×10^{-4}	107	yes
1056	trpC2, metB4, uvr-56	s	+	$<1.0 \times 10^{-5}$	101	yes

a) + indicates wild response; s indicates sensitive response.

b) HCR capacity. Ability to effect host cell reactivation of u.v.-irradiated SP01 phage.

c) From results of alkaline sucrose gradient analysis.

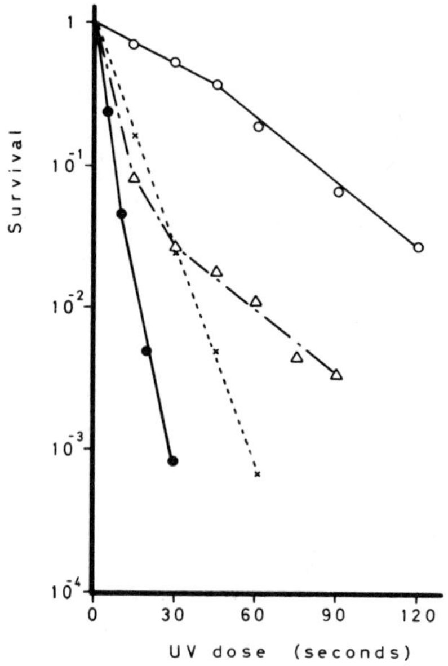

FIG. 1 - SURVIVAL AFTER ULTRAVIOLET IRRADIATION

0 —— 0 GSY228, wild type; x --- x, rec-27; △—— -△, rec-61; ●——●, uvr-56

3. RESULTS

(a) Sensitivity to irradiation

Mutants exr-15, recB19 and rec-61 were originally selected together with a number of other isolates on the basis of their sensitivity to both MMS and MC (Table I). This sensitivity is paralleled by greatly diminished survival after irradiation with ultraviolet light as shown in Figure 1. The survival curves for exr-15 and recB19 are similar to those depicted for rec-27 and rec-61. Also shown is the survival curve for uvr-56, the most u.v.-sensitive strain we have isolated.

The u.v.-survival data are similar to those previously published for the recA1 (Hoch, et al., 1967), recB2 or MC-1 (Okubo and Romig, 1966; Okubo and Nakayama, 1967) and polA (Searashi and Strauss, 1965) recombination defective strains of B. subtilis.

Other data to be reported elsewhere show that these mutants, with the exception of uvr-56, are more sensitive to X-irradiation than the wild type parental strain (Harford, Samojlenko and Mergeay; in preparation). The mutants were also examined for their ability to carry out host cell reactivation (HCR) of u.v.-irradiated SP01 phage and the results of tests with phage irradiated to approximately 7% survival on GSY228 are presented in column 4 of Table I. The suspected rec or exr strains all retain the capacity to effect HCR in contrast to those carrying the uvr-1 and uvr-56 mutations. We conclude that the first group of mutants possesses a functional pathway for the excision of photodimer damage which is defective in the uvr strains.

Figure 2 shows the extent to which prelabelled cellular DNA is degraded to acid soluble fragments upon incubation in growth medium after u.v.-irradiation, at a dose giving 10% survival of wild type.

All three mutants degrade their DNA more extensively than wild type with up to 30% of the total radioactivity being released after 3 hours post-irradiation incubation. The rec-27 culture also shows an appreciable level of spontaneous breakdown in the control. Other data on acid soluble release of DNA after exposure to u.v.-light for strains uvr-1 and uvr-56 indicate that these mutants degrade their DNA to the same or a lesser extent than wild type.

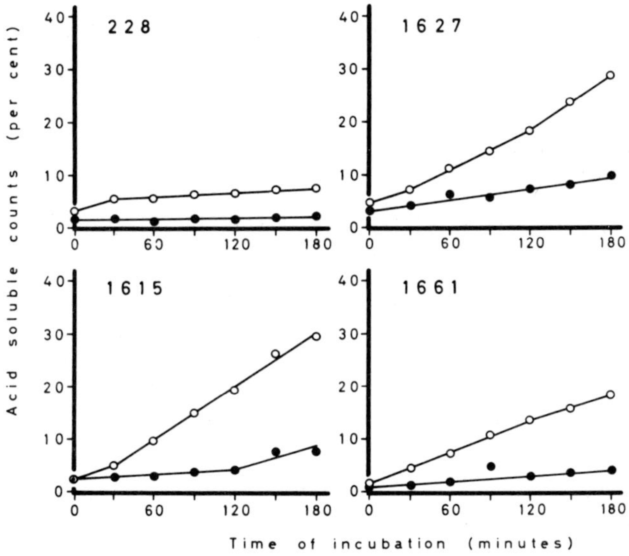

FIG. 2 - BREAKDOWN OF DNA INTO ACID SOLUBLE FRACTION AFTER ULTRAVIOLET IRRADIATION

Cultures were labelled with [^{3}H] thymidine, irradiated for 150 sec and samples taken during reincubation in growth medium for determination of acid soluble counts as described in Materials and Methods. ●——●, Control; O——O, irradiated.

The total cts/min per 0.2 ml. sample and cell survival on TBAB medium after irradiation were as follows:
GSY228, 14700 cts/min; survival 9.9 x 10^{-2}; GSY1615 (exr-15), 6750 cts/min, survival 4.55 x 10^{-4}; GSY1627 (rec-27), 24800 cts/min, survival 5.8 x 10^{-5}; GSY1661 (rec-61), 24900 cts/min, survival 3.3 x 10^{-4}.

Anagnostopoulos and coworkers (unpublished observations) have found that the recA1 strain also displays increased breakdown after u.v.-irradiation. In contrast recB2 strains show low levels of DNA degradation after u.v.-irradiation (Okubo and Nakayama, 1967) as does the recB19 strain listed in Table I.

The recA1, recB2, exr-15, rec-27 and rec-61 markers have been tested for genetic recombination in all possible crosses between donor and recipient and no evidence for linkage was found between any of these loci (Harford, Samojlenko and Mergeay, in preparation). We conclude that each of these mutations is genetically distinct.

The recB19 mutation is strongly linked in genetic recombination tests to recB2 and strains carrying these markers possess identical properties. The exr-15 and exr-18 mutations of the GSY1615 and GSY1618 strains listed in Table I are also closely linked to each other in genetic tests but show no recombination with the other MMS sensitive loci.

(b) Transformation and uptake of ^{3}H-DNA

A suitable test for transformation proficiency is a comparison between the uptake of labelled DNA versus the yield of transformants for a specific marker (Davidoff-Abelson and Dubnau, 1971).

A strain which is blocked in the transformation process following uptake will give a decreased yield of transformants per unit of DNA bound as compared to the wild type and this should be independent of the absolute level of competence for the particular strains tested. The results of a series of tests using a single ^{3}H-DNA preparation are given in Table II and are expressed in the last column as per cent transformability; i.e.

TABLE II

TRANSFORMATION AND UPTAKE OF ^{3}H-DNA BY MUTANTS OF <u>B. SUBTILIS</u>

Recipient	Transformation frequency ($\times 10^{3}$)	<u>trp</u>$^{+}$ transformants per ml ($\times 10^{-5}$)	c.p.m. ^{3}H - DNA taken up per ml [a)]	Ratio <u>trp</u>$^{+}$ transformants / c.p.m. taken up	Transformability relative to wild type (per cent)
GSY228	10.8 8.39	5.47 4.40	4572 3812	120 115 } 117	100
GSY1615 (<u>exr</u>-15)	0.76 0.67	0.15 0.20	331 676	45.4 29.6 } 37.5	32
GSY1618 (<u>exr</u>-18)	1.93	0.53	909	59	50
GSY1619 (<u>recB19</u>)	0.84	0.067	1709	3.9	3.3
GSY1627 (<u>rec</u>-27)	1.13	0.068	270	25.2	21.5
GSY1661 (<u>rec</u>-61)	0.55	0.13	497	26.2	22.4

a) The transformation assay contained 2.83×10^{5} cpm ^{3}H-DNA per ml. of cells.

the ratio of *trp*$^+$ transformants per c.p.m. ^{3}H-DNA taken up, normalized to wild type. The average transformabilities from these and other experiments are given in column 5 of Table I. As shown in Table II the frequency of transformation for each of the mutants is five- to ten-fold lower than that for wild type while the number of *trp*$^+$ transformants/ml. is up to seventyfive-fold lower for *rec*B19 and *rec*-27, reflecting in part the slower growth and reduced viability of these strains.

None of the mutant strains bind as much DNA as wild type indicating lower levels of competence in these cultures.

The normalized ratios for *trp*$^+$ transformants to c.p.m. ^{3}H-DNA taken up indicate that *rec*B19 (and *rec*B2) retain about 5%, and *rec*-27 and *rec*-61 about 20%, wild type transformability. We therefore judge these strains to be partially blocked in some step in the transformation process following DNA uptake. The results for strains GSY1615 and GSY1618 (*exr*-15, -18) are more inconclusive as they retain up to 50% transformability although poorly competent. For this reason and on the basis of other properties we have tentatively assigned these mutants an *exr* designation by analogy with the EXR phenotype of *E. coli* (Rorsch *et al*., 1966). In other experiments we have found that *uvr* strains possess full transformability in agreement with the previous data of Hoch *et al*. (1967) based on transformation frequencies.

(c) Repair of ultraviolet light induced damage to DNA

Each of the mutants has been examined by the alkaline sucrose gradient technique of Mc Grath and Williams (1966) for its capacity to repair single strand breaks induced in prelabelled DNA after u.v.-irradiation. Since the cells are prelabelled it is important to emphasize that we are not directly investigating the ability of the mutants to perform post-replication repair (Rupp and Howard-Flanders, 1968; Smith and Meun, 1970). The results are presented in Figures 3 to 9 as the per cent distribution of total radioactivity recovered from the gradients. In all experiments control cells of GSY228 labelled with $[^{14}C]$ thymidine were added to each gradient to ensure that gradient conditions and the position of the fast sedimenting control peak did not vary within an experiment.

The ability of wild type cells of B. subtilis to effect repair of u.v.-damage to DNA is shown in Figure 3. Immediately after irradiation the fast sedimenting DNA peak in the control is converted to a slower sedimenting, lower molecular weight form due to the presence of single strand interruptions presumably induced by the excision of pyrimidine dimers (Setlow and Carrier, 1964; Boyce and Howard-Flanders, 1964). During post-irradiation incubation this damaged DNA is slowly converted to faster sedimenting material. After 90 to 100 minutes incubation, it sediments in the same position as the control peak indicating the action of repair systems for the rejoining of single strand breaks (see Howard-Flanders, 1968; Smith, 1971; for reviews). Molecular weight values for the various DNA species represented in these experiments have not yet been investigated. Under similar experimental conditions Tanooka and Terano (1970) estimated the single strand average molecular weight of the DNA from both spores and vegetative cells of B. subtilis to be 1.1×10^8 daltons.

In contrast to the results for wild type, cells of rec-27 (Figure 4) and exr-15 (Figure 5) show little evidence of repair capacity during post-irradiation incubation although a small amount of DNA sedimenting near the peak position is evident in exr-15 after 30 minutes incubation. No increase in the sedimentation rate of DNA from either mutant was detected after 100 minutes incubation.

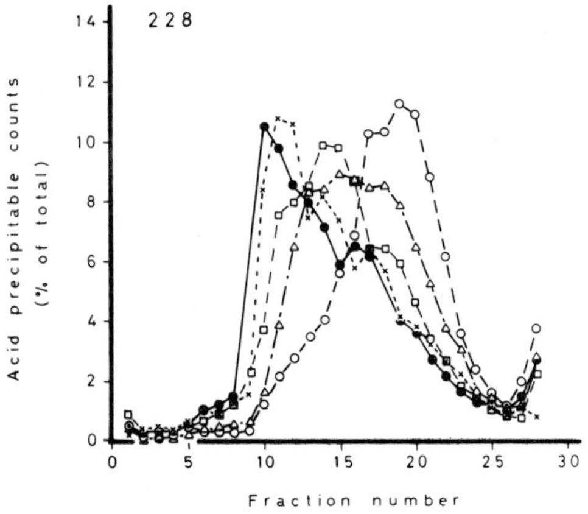

FIG. 3 - EFFECT OF REINCUBATION ON SEDIMENTATION IN ALKALINE SUCROSE OF PRELABELLED DNA FROM U.V.-IRRADIATED CELLS OF GSY228 (rec$^+$, uvr$^+$).

Cultures were prelabelled with $[^3H]$ thymidine, irradiated with ultraviolet light for 150 sec and reincubated as described in Materials and Methods. Cell survival on TBAB medium was not measured in this experiment but in other experiments under the same conditions the survival was 7% to 11%.

●——● control, no irradiation (100% = 15090 cts/min);
O——O irradiated, no reincubation (100% = 17390 cts/min);
△——△ irradiated, 30 min reincubation (100% = 16240 cts/min);
□——□ irradiated, 60 min reincubation (100% = 17580 cts/min);
x----x irradiated, 100 min reincubation (100% = 16870 cts/min).

Both mutant rec-61 (Figure 6) and mutant recB19 (Figure 7) retain a significant capacity to convert u.v.-damaged DNA to a faster sedimenting form during post-irradiation incubation in contrast to the results for rec-27 and exr-15. The distribution of radioactivity in the 100 min. sample from rec-61 cells is similar to that found for the control although cell survival in this strain is over one hundred-fold lower than the survival for wild type at the same u.v.-dose. RecB19 shows a lesser capacity for repair with approximately 40% of the total radioactivity cosedimenting at the control peak position after 100 minutes incubation.

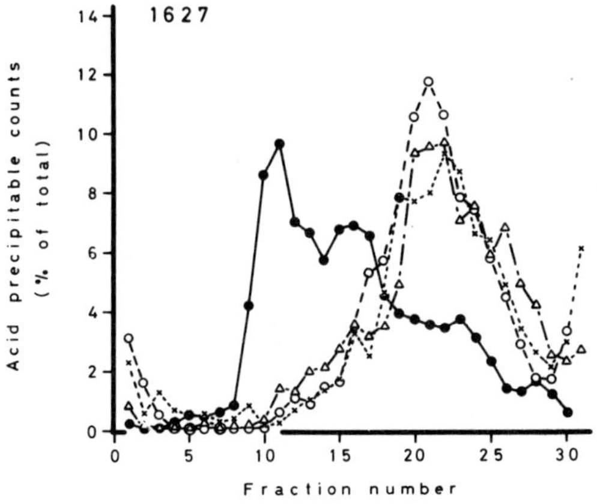

FIG. 4 - EFFECT OF REINCUBATION ON SEDIMENTATION IN ALKALINE SUCROSE OF PRELABELLED ^{3}H-DNA FROM U.V.-IRRADIATED CELLS OF GSY1627 (rec-27).

Experimental conditions were as described in Materials and Methods. Cell survival after irradiation was 2.56×10^{-4}.

●——● , Control, no irradiation (100% = 3046 cts/min);
O——O , irradiated, no reincubation (100% = 4011 cts/min);
△——△ , irradiated, 30 min. reincubation (100% = 2844 cts/min);
x----x , irradiated, 100 min. reincubation (100%= 4006 cts/min).

The u.v.-sensitive mutants uvr-56 and uvr-1 were also examined in similar experiments with the results shown in Figures 8 and 9. In both strains there is a conversion of DNA to a lower molecular weight after irradiation but the extent of this shift is less than in the other strains as 25 to 30% of the total radioactivity still sediments in the region of the control peak. Both mutants retain some capacity for repair after post-irradiation incubation although an appreciable fraction of the radioactivity in the 100 minute sample from uvr-1 (Figure 9) sediments at the bottom of the gradient. This is probably due to incomplete lysis of cells on the gradient.

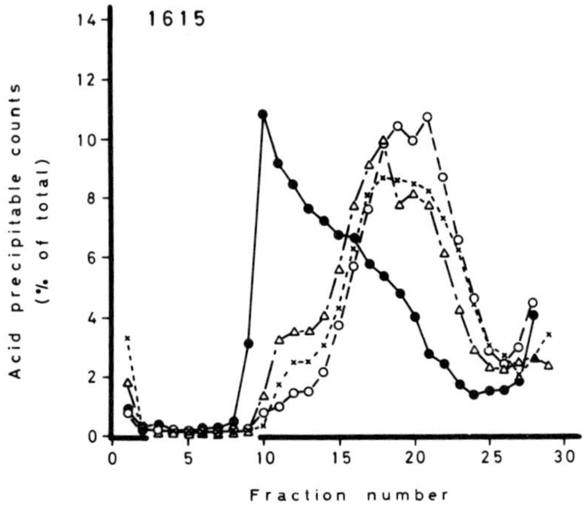

FIG. 5 - EFFECT OF REINCUBATION ON SEDIMENTATION IN ALKALINE SUCROSE OF PRELABELLED ^{3}H-DNA FROM U.V.-IRRADIATED CELLS OF GSY1615 (exr-15).

Experimental conditions were as described in Materials and Methods. Cell survival after irradiation was 2.2 x 10^{-5}.

●——● , Control, no irradiation (100% = 12700 cts/min);
O——O , irradiated, no reincubation (100% = 12160 cts/min);
△——△ , irradiated,30 min. reincubation (100%=9979 cts/min);
x----x , irradiated,90 min. reincubation (100%=9155 cts/min).

A complicating factor in experiments of this type is the influence of the considerable degradation of DNA in the rec-27, rec-61 and exr-15 mutants. Absence of observable repair after irradiation may be due either to a block in the repolymeri

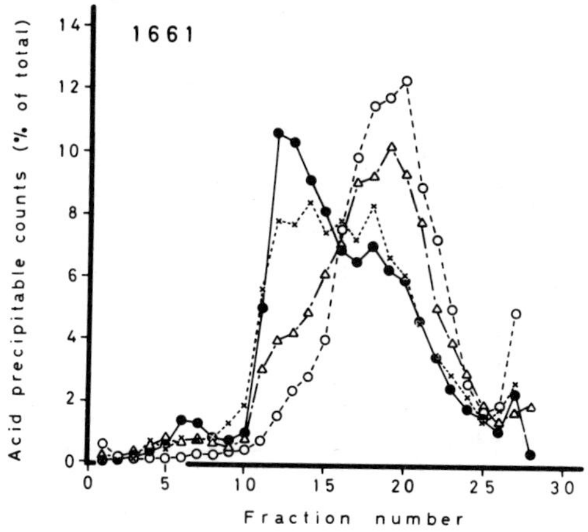

FIG. 6 - EFFECT OF REINCUBATION ON SEDIMENTATION IN ALKALINE SUCROSE OF PRELABELLED ^{3}H-DNA FROM U.V.-IRRADIATED CELLS OF GSY1661 (rec-61).

Experimental conditions were as described in Materials and Methods. Cell survival after irradiation was 1.59×10^{-4}.

●——●, Control, no irradiation (100% = 22630 cts/min);
O——O, irradiated, no reincubation (100%=19840 cts/min);
Δ——Δ, irradiated, 30 min. reincubation (100%=20460 cts/min);
x----x, irradiated, 100 min. reincubation (100%=17320 cts/min).

zation and sealing of single strand regions or to the supervention of an uncontrolled nuclease which attacks the sites at which genetic damage is being repaired and therefore masks or destroys any potential repair (Clark, Chamberlin, Boyce and Howard-Flanders, 1966). Degradation of newly synthesized DNA in recA cells of *E. coli* was found to be reduced by glucose and amino acid starvation prior to irradiation but even under these conditions no evidence for post-replication repair could be obtained (Smith and Meun, 1970). The evidence for repair in rec-61 indicates that this process can be observed in cells which degrade up to 20% of their DNA to acid soluble fragments (Figure 2).

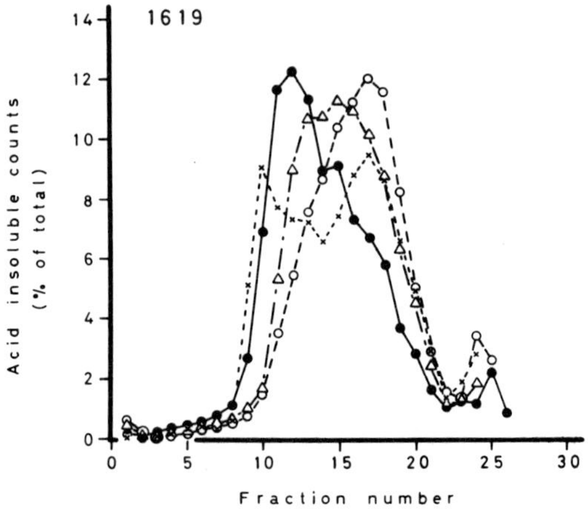

FIG. 7 - EFFECT OF REINCUBATION ON SEDIMENTATION IN ALKALINE SUCROSE OF PRELABELLED ^{3}H-DNA FROM U.V.-IRRADIATED CELLS OF GSY1619 (recB19).

Experimental conditions were as described in Materials and Methods. Cell survival after irradiation was 8.10 x 10^{-4}.

●——● , Control, no irradiation (100% = 26233 cts/min);
0——0 , irradiated, no reincubation (100%=24304 cts/min);
Δ——Δ , irradiated, 30 min reincubation (100%=26924 cts/min);
x----x , irradiated, 100 min reincubation(100%=22330 cts/min).

4. DISCUSSION

The recB19, rec-27 and rec-61 mutants whose properties are summarized in Table I belong to the class of transformation defective mutants which are radiation sensitive and blocked in some step of the recombination process following the uptake of DNA into the competent cell. The recB19 mutant is found to retain about 5% and the rec-27 and rec-61 mutants up to 20% of wild type transformability indicating either that the mutations are leaky or that alternative pathways exist for processing recombinant molecules in the B. subtilis transformation system.

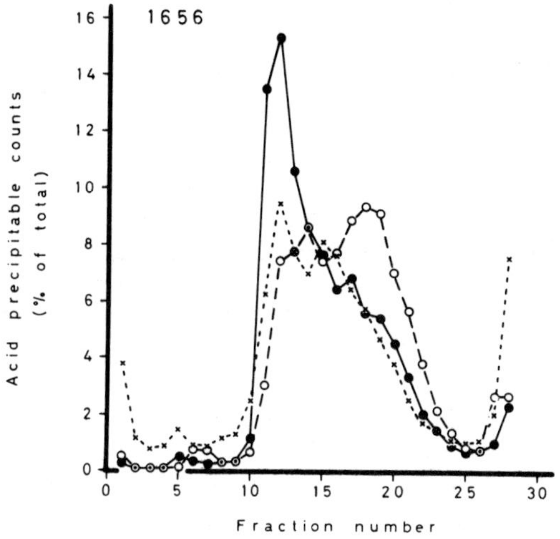

FIG. 8 - EFFECT OF REINCUBATION ON SEDIMENTATION IN ALKALINE SUCROSE OF PRELABELLED ^{3}H-DNA FROM U.V.-IRRADIATED CELLS OF GSY1656 (uvr-56).

Experimental conditions were as described in Materials and Methods. Cell survival after irradiation was 2.6×10^{-7}.

●——●, Control, no irradiation (100% = 11190 cts/min);
O——O, irradiated, no reincubation (100% = 11980 cts/min);
x----x, irradiated, 90 min. reincubation (100%=9933 cts/min).

Hoch et al. (1967) and Hoch and Anagnostopoulos (1970) found that transformation and SP10 transduction, but not transduction by PBS1, were impaired in the recA1 strain and suggested that (i) alternative pathways do exist, at least for differing types of donor molecules, or (ii) the step controlled by the recA1^{+} gene is not essential for PBS1 transduction. Further evidence for two modes of recombination was obtained by Dubnau, Davidoff-Abelson and Smith (1969) who found that both recA1 and

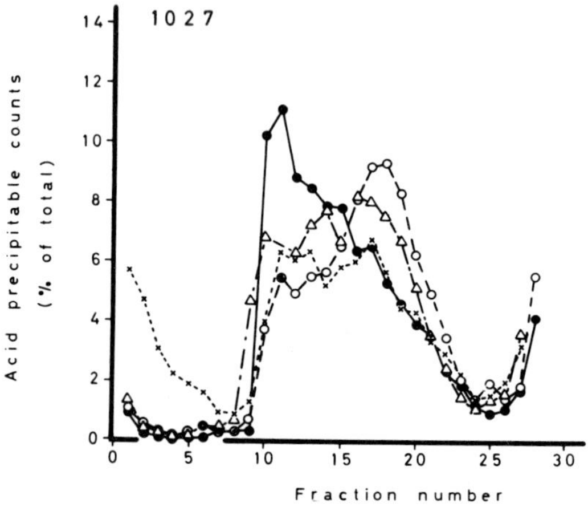

FIG. 9 - EFFECT OF REINCUBATION ON SEDIMENTATION IN ALKALINE SUCROSE OF PRELABELLED ^{3}H-DNA FROM U.V.-IRRADIATED CELLS OF GSY1027 (uvr-1).

Experimental conditions were as described in Materials and Methods. Cell survival after u.v.-irradiation was not measured in this experiment but in other experiments under the same conditions survival was 1.17×10^{-3}.

●——●, Control, no irradiation (100% = 11042 cts/min);
O——O, irradiated, no reincubation (100% = 13100 cts/min);
Δ——Δ, irradiated, 30 min reincubation (100%=19520 cts/min);
x----x, irradiated, 100 min reincubation(100%=15290 cts/min).

wild type 168 strains were transduced at lowered frequencies by PBS1 phage grown on the partially heterologous W23 donor strain. They suggested that one pathway (controlled by *recA*$^{+}$) deals with transformation and is tolerant towards heterologous transducing molecules; the second pathway is predominantly for homologous transduction. Hoch and Anagnostopoulos (1970) constructed a doubly mutant *recA*1, *recB*2 strain, but found this to be no more defective for SP10 transduction than the single *recA*1 mutant. It was not possible to decide from this data whether the effects of the two mutations were additive or whether the effect of *recA1* was predominant. Further tests and the construction of other double mutants are necessary to determine if given *rec* mutants contribute to the same or differing pathways. Two alternative pathways are known to be involved in recombination in *E. coli* (Clark, 1971).

A third series of mutants isolated in our laboratory have been classified as EXR (Table I). Although these mutants are radiation sensitive and share properties in common with *recA1*, *rec*-27 and *rec*-61, including increased breakdown of DNA after u.v.-irradiation and show reduced competence, they retain up to 50% transformability. The exact phenotypic and genotypic properties of this locus must be determined by further studies.

The mutants are defined here as recombination deficient on the basis of reduced transformability and we have not yet fully investigated other processes such as transfection and transduction in which genetic recombination is involved. During transfection a primary, host mediated genetic recombination step is obligatory for those phages with a multipowered dose relationship between DNA concentration and output of infective centres (Spatz and Trautner, 1971). All recombination defective mutants are not necessarily deficient in this required step or steps for transfection. Transfection by SP01 DNA has been found to be greatly reduced or absent in the *recB2* strain while SP02 transfection in unaffected (Okubo and Romig, 1966). A preliminary test with DNA from phage SP01 (Okubo, Strauss and Stodolsky, 1964) indicated that for *rec*-61 the ratio of transfectants to *trp*$^+$ transformants is the same as for wild type; i.e. transfection in *rec*-61 is reduced by the same extent as transformation.

With rec-27 the ratio of transfectants to transformants is ten to thirtyfold higher than the same ratio for wild type indicating that this strain is normal or proficient for transfection. The absolute frequency of transfectants per viable cell was also found to be higher in rec-27 than in wild type. The interpretation of these results for rec-27 is made more complex by the fact that this strain shows a four to five-fold lower viability than wild type at the same optical density although protein and DNA content are approximately equivalent. Preliminary electron microscope studies of rec-27 carried out in this laboratory by Mr. D.Charles reveal no obvious structural abnormalities in cells of this mutant.

Either the defect in rec-27 does not affect the primary recombination step in transfection or the "inviable" cell fraction may be incapable of producing transformed progeny but permit transfection since this does not involve cell survival.

At present we have no information on the nature of the enzyme defects in these mutants. Dr. Janine Doly has tested the mutants for DNA polymerase I and ATP stimulated deoxyribonuclease activities and found normal levels of both enzymes (J. Doly and C. Anagnostopoulos, personal communication).

Our results with alkaline sucrose gradient analysis of prelabelled DNA from u.v.-irradiated cells show that the wild type, recB19 and rec-61 strains are capable of repair of single strand break damage whereas rec-27 and exr-15 are not. The two uvr mutants examined are also capable of some repair although the number of breaks induced after irradiation appears to be less than in excision proficient strains. The present experiments differ from those of Rupp and Howard-Flanders (1968) and

Smith and Meun (1970) in that we are not directly examining the repair of gaps present in the DNA sythesized immediately after irradiation on a damaged template, but in preexisting DNA. Excision defective mutants of E. coli are capable of an independent type of repair of u.v.-damage to DNA, following chromosome replication, which involves exchanges between sister DNA duplexes to form an undamaged copy. This process has been termed recombinational repair (Rupp and Howard-Flandres, 1968; Howard-Flanders, 1968; Rupp, Wilde, Reno and Howard-Flanders, 1971). The exchanges resemble those involved in genetic recombination and are directed by some of the same enzymes since the $recA^{+}$ gene product is necessary for this type of repair. Wild type, uvr, recB and recC strains, but not recA, effect post-replication recombinational repair in E. coli (Smith and Meun, 1970).

In the experiments presented in Figures 3 to 9 it is not possible to assess the role of post-irradiation chromosome replication in the repair observed for wild type, recB19 and rec-61; i.e. the relative contributions of pre- and post-replication repair cannot be distinguished. However it is clear that rec-61 and recB19 retain functional mechanisms for the polymerization and sealing of single strand gaps induced by the excision of u.v. lesions. We presume that, by analogy with E. coli, the repair observed in the uvr strains is the result of recombinational repair mechanisms acting after post-irradiation DNA synthesis since these cells are probably unable to perform dimer excision and are proficient in transformation. The influence of DNA replication on repair after irradiation will become directly testable when the present mutants are introduced into a thymine requiring strain. Both the rec-27 and exr-15 strains retain an unimpaired mechanism for photodimer excision as judged by their ability to effect HCR and by the number of breaks intro duced into their DNA after u.v.-irradiation. However neither strain appears capable of repair on subsequent incubation although both mutants have been found to possess normal levels of DNA polymerase I activity. As noted previously it is not clear whether absence of repair is due to the widening and destruction of potential repair sites by nuclease action or to a block in the repair process per se.

The mutant strains have also been investigated for the ability to repair damage to their DNA after X-ray irradiation. The responses are similar to those seen after u.v.-irradiation; recB19 and rec-61 retain some capacity for the rejoining of single strand breaks while the rec-27 and exr mutants do not (Samojlenko and Harford, unpublished observations). We conclude that repair of single strand break damage to DNA directly requires the exr-15$^+$ and rec-27$^+$ gene products. The loci represented by the recB19 and rec-61 mutations must also be involved in survival after irradiation since strains carrying these markers are radiation sensitive. However it is not possible at present to assess which steps in the repair processe are controlled by these genes.

Dubnau and co-workers have recently made considerable progress in the detailed analysis of the conversion of donor DNA to recombinant molecules in the *B. subtilis* transformation system (Dubnau and Cirigliano, 1972). However only two rec mutants have been studied in any detail to detect their recombinational defect. One mutant appears unable to form covalent bonds between the synapsed donor and resident DNA molecules(Zadrazil and Fucik, 1971). Recombination in the recB2 strain has been studied by Davidoff-Abelson and Dubnau (1971) who found that integration of donor DNA proceeded normally but that potential transformants were lost during post-integration chromosome replication. It is of interest that caffeine has a selective killing effect on potential transformants until about the time when multiplication of these cells can be detected in controls (Harford and Legge, unpublished observations). Studies on the fate of both homologous and heterologous donor DNA in our mutants are now in progress and should assist in elucidating the sequence of molecular events and the enzyme mechanisms involved in genetic recombination and repair in *B. subtilis*.

ACKNOWLEDGEMENTS

We wish to express our gratitude to Prof. L. Ledoux for his support of this work. We also wish to thank Dr. C. Anagnostopoulos for helpful discussions, permission to quote unpublished results and for the gift of bacterial strains. During the course of this work I. S. was supported by a Cultural Exchange Programme Fellowship from the Belgian National Ministry of Education.

5. REFERENCES

ANAGNOSTOPOULOS, C. & SPIZIZEN, J. (1961). J. Bact. 81:741.

BEATTIE, K.L. & SETLOW, J.K. (1971). Nature New Biol. 231:177.

BOYCE, R. & HOWARD-FLANDERS, P. (1964). Proc. Nat. Acad. Sci., Wash. 51:293.

CLARK, A.J. (1971). Ann. Rev. Microbiol. 25:437.

CLARK, A.J., CHAMBERLIN, M., BOYCE, R.P. & HOWARD-FLANDERS, P. (1966). J. Mol. Biol. 19:442.

CLARK, A.J. & MARGULIES, A.D. (1965). Proc. Nat. Acad. Sci., Wash. 53:451.

DAVIDOFF-ABELSON, R. & DUBNAU, D. (1971). Proc. Nat. Acad. Sci., Wash. 68:1070.

DUBNAU, D. & CIRIGLIANO, C. (1972). J. Mol. Biol. 64:31.

DUBNAU, D., DAVIDOFF-ABELSON, R. & SMITH, I. (1969). J. Mol.Biol. 45:155.

FARMER, J.L. & ROTHMAN, F. (1965). J. Bact. 89:262.

HARFORD, N. & SUEOKA, N. (1970). J. Mol. Biol. 51:267.

HILL, T., PRAKASH, L. & STRAUSS, B. (1972). J. Bact. 110:47.

HOCH, J.A. & ANAGNOSTOPOULOS, C. (1970). J. Bact. 103:295.

HOCH, J.A., BARAT, M. & ANAGNOSTOPOULOS, C. (1967). J. Bact. 93: 1925.

HOLLOWAY, B. (1966). Mutation Res. 3:452.

HOWARD-FLANDERS, P. (1968). Ann. Rev. Biochem. 37:175.

HOWARD-FLANDERS, P. & BOYCE, R.P. (1966). Radiation Res. (Suppl. 6) P. 156.

KOOISTRA, J. & VENEMA, Gs. (1970). Mutation Res. 9:245.

MC GRATH, R. & WILLIAMS, R.W. (1966). Nature, 212:534.

OKUBO, S. & NAKAYAMA, H. (1967). Mutation Res. 4:533.

OKUBO, S. & ROMIG, W.R. (1966). J. Mol. Biol. 15:440.

OKUBO, S., STRAUSS, B. & STODOLSKY, M. (1964). Virology, 24:552.

POSTEL, E.H. & GOODGAL, S.H. (1972). J. Bact. 109:292.

REITER, H. & STRAUSS, B. (1965). J. Mol. Biol. 14:179.

RORSCH, A., VAN DE PUTTE, P., MATTERN, I., ZWENK, H. & VAN SLUIS, C. (1966). In: Genetical aspects of radiosensitivity: Mechanisms of repair. Vienna: Int. Atomic Energy Agency.

RUPP, W.D. & HOWARD-FLANDERS, P. (1968). J. Mol. Biol. 31:291.

RUPP,W.D., WILDE, C.E., RENO, D.L. & HOWARD-FLANDERS, P. (1971). J. Mol. Biol. 61:25.

SEARASHI, T. & STRAUSS, B. (1965). Biochem. Biophys. Res. Comm. 20:680.

SETLOW, R. & CARRIER, W. (1964). Proc. Nat. Acad. Sci.,Wash. 51:226.

SMITH, K.C. (1971). In Photophysiology, vol. VI pp. 210-278. A.C. Giese (Ed.). New York: Academic Press.

SMITH, K.C. & MEUN. D.H.C. (1970). J. Mol. Biol. 51:459.

SPATZ, C. & TRAUTNER, T.A. (1971). Molec. Gen. Genetics, 113:174.

TANOOKA, H. & TERANO, H. (1970). Radiation Res. 43:613.

ZADRAZIL, S. & FUCIK, V. (1971). Biochem. Biophys. Res. Comm. 42:676.

INTERGENOTIC AND HETEROSPECIFIC TRANSFORMATION: THE MECHANISM OF RESTRICTION OF GENETIC EXCHANGE IN *BACILLUS SUBTILIS*

GARY A. WILSON

FRANK E. YOUNG

The University of Rochester School of Medicine and Dentistry
Rochester, New York 14642, U.S.A.

ABSTRACT

Transformation of *Bacillus subtilis* 168 by DNA of other bacilli occurs at a reduced frequency. The mechanism of this reduction was studied by constructing strains of *B. subtilis* 168 and *B.subtilis* HSR which were lysogenic for the bacteriophage Ø105. This provided a marker in the heterologous DNA from *B. subtilis* HSR with a nucleotide sequence identical to the one in the recipient strain. Using defective and clear plaque mutants of Ø105 it was possible to investigate genetic exchanges which were analogous to transformation with heterologous, homologous and intergenotic DNA by varying the neighborhood of the DNA surrounding the prophage. Transfection of *B.subtilis* 168 with DNA from the prophage was inefficient when Ø105 was integrated in the heterologous strain(*B.subtilis* HSR) This restricted exchange in *B.subtilis* 168 was partially relieved when the source of DNA was from the progeny of *B. subtilis* HSR (Ø105). Transfection in this case corresponds to intergenotic transformation. Recombination between prophage DNA and a recipient strain carrying a defective prophage was equally efficient regardless of the neighborhood of DNA surrounding the prophage in the donor DNA.Therefore, classical modification and restriction in the region of the intergenote is not a plausible explanation for the decreased efficiency of transfection by heterologous DNA. These results suggest that the restriction of genetic exchange with heterologous DNA and the subsequent increased efficiency of transformation with intergenotic DNA is dependent on the relative degree of homology of nucleotide sequences in the region of recombination.

1. INTRODUCTION

Genetic exchange between B. subtilis 168 and other bacilli is generally very inefficient and occurs at a measurable frequency for only a limited number of loci. These loci predominate in two regions of the chromosome and are thought to represent areas which have been conserved(Chilton & McCarthy,1969; Dubnau, Smith, Morell & Marmur, 1965). This genetic conservation suggests that the nucleotide sequences of these regions, unlike the rest of the chromosome,have not changed significantly during the divergence of the bacillus genospecies(Ravin, 1963). Genetic exchange, therefore,should be efficient between organisms in any region of identical or nearly identical nucleotide sequence. Genetic hybrids (Biswas & Ravin, 1971) or intergenotes (Wilson & Young, 1972) have been constructed to investigate exchange by heterologous DNA.Once the segment of foreign or heterologous DNA has been integrated in the neighborhood of homologous DNA, the frequency of transformation with this intergenotic DNA should then be equal to that of homologous DNA. However, recent genetic studies have shown that intergenotic DNA, although markedly more efficient than heterologous DNA, is rarely as efficient a transforming agent as is homologous DNA (Wilson & Young, 1972).

In previous studies,it has not been possible to determine if the inefficiency or restriction was due to small differences between the nucleotide sequence in the intergenotic region of the donor and the recipient or if modification(and subsequent restriction) of the foreign integrated segment of DNA was responsible for the decreased efficiency. By integrating the temperate bacteriophage Ø105 into the genome of the heterologous strain *B. subtilis* HSR, we constructed heterologous DNA containing a genetic marker with a nucleotide sequence identical to that found in the homologous strain.Transformation with DNA from these strains demonstrates that the inefficient transformation by intergenotic DNA is related to the nucleotide sequence of the genome and not to modification of the intergenotic region.

2. MATERIALS AND METHODS

(a) Bacterial strains and bacteriophage

The bacterial strains and their related bacteriophage are summarized in Table I.This table does not include the strains of *B.pumilus*, *B.licheniformis*, *B.amyloliquefaciens* or *B. subtilis*, which were described previously (Wilson & Young, 1972) and were used (Table II and III)to summarize genetic studies which resulted in the model presented in Figure 1.

(b) Isolation of lysogenic strains

In order to isolate *B.subtilis* HSR(Ø105),bacteriophage Ø105 was spotted on a lawn of *B. subtilis* HSR in semisolid agar (Yásbin & Young, 1972). After overnight incubation the infected area was streaked on solid agar seeded with bacteriophage Ø105 clz. This clear plaque mutant of Ø105 cannot successfully infect

TABLE I

BACTERIOPHAGE AND STRAINS OF BACILLI

Strain of bacteriophage	Genotype or phenotype	Source and comments (references)
B.subtilis BR151	*lys*-3,*trp*C2,*met*B10	B. E. Reilly
B.subtilis BR151(Ø105)†	*lys*-3,*trp*C2,*met*B10(Ø105)	lysogenic strain of BR151 by infection with Ø105
B.subtilis 173	*his* B, *met* B, *sup*-3	M. J. Tevethia‡
B.subtilis HSR	Str^R , Trp^-	B.E. Reilly,Streptomycin resistant strain of *B.subtilis* H (Reilly & Spizizen, 1965)
B.subtilis HSR(Ø105)	Str^R , Trp^- (Ø105)	lysogenic strain of HSR by infection with Ø105
B.subtilis RUB300	*lys*-3, *his*B2, *met*B10	*his*B2 introduced into BR151 by congression (Nester, *et al*., 1963)
B.subtilis RUB310	*lys*-3,*trp*C2,*met*B10(Ø105)	lysogenic strain of BR151 by transformation with DNA from BR151 (Ø105)
B.subtilis RUB311	*lys*-3,*trp*C2,*met*B10(Ø105 *sus* 11)	lysogenic strain of BR151 by infection with Ø105 *sus* 11
B.subtilis RUB316	*lys*-3,*trp*C2,*met*B10(Ø105 *sus* 19)	lysogenic strain of BR151 by infection with Ø105 *sus* 19
B.subtilis RUB407	*lys*-3,*trp*C2,*met*B10(HSR Ø105)§	lysogenic strain of BR151 by transformation with prophage DNA from HSR (Ø105)

B.subtilis RUB408	*lys*-3,*trp*C2,*met*B10(HSR Ø105)§	see RUB407
B.subtilis RUB409	*lys*-3,*trp*C2,*met*B10(HSR Ø105)§	see RUB407
B.subtilis RUB410	*lys*-3,*trp*C2,*met*B10(HSR Ø105)§	see RUB407
B.subtilis RUB511	Str^R, Trp^- (Ø105 *sus* 11)	lysogenic strain of HSR by infection with Ø105 *sus* 11
B.subtilis RUB516	Str^R, Trp^- (Ø105 *sus* 19)	lysogenic strain of HSR by infection with Ø105 *sus* 19
B.subtilis RUB518	Str^R, Trp^- (Ø105 *sus* 19)	see RUB516
Ø105	wild type	B. E. Reilly
Ø105 *sus* 11	*sus* 11	L.Rutberg(Rutberg & Armentrout, 1970)
Ø105 *sus* 19	*sus* 19	L.Rutberg(Rutberg & Armentrout, 1970)
Ø105 clz	clear plaque mutant	S. Zahler

† (Ø105) indicates that the strain is lysogenic for bacteriophage Ø105.

‡ A strain which is isogenic to BR151 was transformed with DNA containing the suppressor-3 mutation (Georgopoules, 1969). The *his* B mutation is suppressed in this strain.

§ (HSR Ø105) indicates that the strain is lysogenic for bacteriophage Ø105 and may contain some heterologous DNA from *B. subtilis* HSR.

TABLE II

EFFICIENCY OF TRANSFORMATION OF BR151 WITH HOMOLOGOUS AND HETEROLOGOUS DNA

Donor DNA		Relative frequency of transformation			
		Lys^+	Trp^+	Met^+	Rfm^R
B. subtilis	homologous	1.0	1.0	1.0	1.0
B. pumilus	heterologous	0.006	0.002	0.0001	0.003
B. licheniformis	heterologous	0.0003	0.0003	0.0002	0.002
B. amyloliquefaciens H	heterologous	0.0002	0.0005	0.0002	0.12
B. subtilis HSR	heterologous	0.0003	ND†	0.0001	0.10

† Not determined

TABLE III

EFFICIENCY OF TRANSFORMATION OF RUB300 WITH INTERGENOTIC DNA

Origin of intergenotic marker	Relative frequency of transformation		
	HisB†	Lys^+	Rfm^R
B.pumilus	0.91	0.49	0.54
B.licheniformis	1.0	ND‡	0.20
B.amyloliquefaciens H	1.0	0.24	2.57
B.subtilis HSR	1.0	ND	0.51

† Homologous control marker

‡ Not determined

cells which are lysogenic for Ø105. Clones were reisolated that were resistant to Ø105 and capable of releasing only wild type Ø105, thus indicating that the resistant cells were not carrying the clear plaque mutant that was used in the selection process. The DNA isolated from these strains was capable of low levels of transfection and was markedly restricted in its ability to transform auxotrophic mutants of *B. subtilis* 168.

The lysogenic strains carrying the intergenotic prophage region(RUB 407, RUB 408, RUB 409,and RUB 410)were isolated by transforming *B. subtilis* BR151 with DNA isolated from *B.subtilis* HSR (Ø105). The transformed cells were plated in semisolid agar and incubated for three hours at 37° C. These plates were then challenged with 2.0 ml semisolid agar seeded with bacterio-

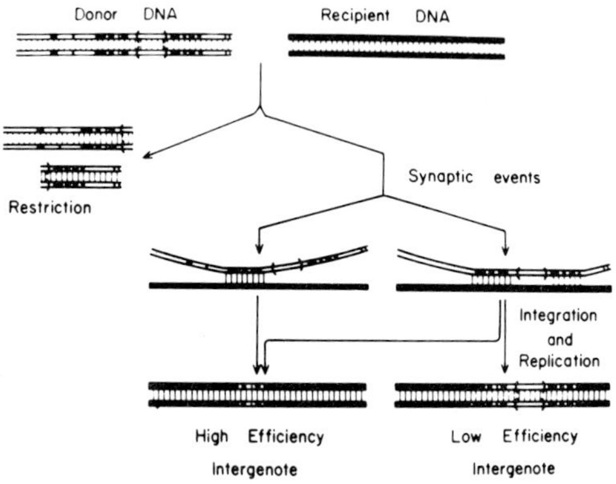

Fig. 1 - The fate of heterologous transforming DNA. Double-stranded DNA is represented with two heavy bars. Regions of identical base sequences are solid black. Light vertical lines indicate proper base pairing between the two strands of DNA. The area enclosed by parentheses is the site of recognition for a restrictive enzyme. Reprinted with permission from the American Society for Microbiology.

phage Ø105 clz. Clones were selected which remained resistant to Ø105 clz and released wild type bacteriophage Ø105. DNA isolated from these strains could transfect and transform auxotrophic mutants of B. subtilis 168 demonstrating that the wild type Ø105 had been integrated in the genome, and that the majority of the DNA was from the chromosome of the recipient.

The lysogenic strains carrying suppressor mutants of Ø105 were obtained by infecting B. subtilis BR151 or HSR with Ø105 sus 11 or Ø105 sus 19. Colonies which were resistant to Ø105 clz were isolated as described previously. These strains remained resistant to Ø105 clz and were able to produce plaques on B. subtilis 173 but not on B. subtilis 168 which lacked the suppressor gene.DNA isolated from these strains could transfect B. subtilis 168 and recombine with the recipient to produce wild type Ø105 plaques.

(c) Isolation of DNA and growth of cells to competence

The procedures for phenol extraction of transforming DNA and growth of competent cells in either a single medium or a stepdown procedure have been described elsewhere (Wilson & Young, 1972).

(d) Assay of biological activity

After exposure of competent cells to DNA for a period of 8 to 30 min, 50 μg of pancreatic deoxyribonuclease (DNase) was added to terminate transformation or transfection. Cells were diluted in Spizizen's minimal medium (Spizizen, 1958) and plated on minimal agar in 2.0 ml semisolid medium(Yasbin & Young, 1972) containing 0.5% casein acid hydrolysate and 400 μg tryptophan. Indicator bacteria (either B. subtilis 173 or BR151) were added to determine plaque forming units. To determine resistance to bacteriophage Ø105 clz, cells were incubated for three hours at 37° C after plating in semisolid medium before being challenged with Ø105 clz in an additional 2.0 ml of semisolid agar overlay.

3. RESULTS

(a) Efficiency of transformation with homologous and heterologous DNA

To elucidate the nature of the restriction of genetic exchange between closely related bacilli,we decided to limit the investigation to B.subtilis,B.pumilus,B.licheniformis,B.amyloliquefaciens H, and B. subtilis HSR. These bacilli are related as indicated by the similarities in the GC content of the DNA (Lovett & Young,1969;Young & Wilson,1972).Although DNA-DNA hybridization between species varied from only 11 to 24% (Lovett & Young,1969) (the hybridization data for B. subtilis HSR are not available) genetic exchange between species has been possible by DNA mediated transformation(Wilson & Young,1972).The efficiency of transformation is summarized in Table II. Although transformation for the auxotrophic loci, lys-3, trpC2 and metB10, was markedly inefficient, transformation to rifampin resistance occurred at a higher frequency. Because of the increased efficiency for this marker, it was possible to isolate strains of B. subtilis which had been transformed by heterologous DNA.These strains are called intergenotes because they contain an integrated fragment of a foreign genome.As shown in Table III, transformation with intergenotic DNA was much more efficient than it was with heterologous DNA, but in most cases, the efficiency was not equal to that observed with homologous DNA.These results have been detailed previously(Wilson & Young,1972)and are summarized here only because they resulted in the development of a model(Figure 1) to explain restriction of heterologous DNA in transformation. Because this communication emphasizes experiments which were designed to specifically test one aspect of the model,only the salient features of the model will be discussed here.

When heterologous DNA enters a cell, genetic exchange is restricted by endonucleases which recognize a specific nucleotide sequence and cut the DNA into pieces which are below the optimal length for integration. Restriction enzymes with these properties have been described (Gromkova & Goodgal,1972; Kelly & Smith,1970; Middleton, Marshall & Hutchison,1972; Smith & Wilcox, 1970).When heterologous DNA does successfully integrate, the recognition site for the endonuclease is excised and the efficiency of transformation should now be dependent only on the effectiveness of the base pairing in the region of the high efficiency intergenote.As shown in the model,the nucleotide sequence in the intergenotic region may still be slightly different than the homologous DNA thus accounting for the slightly reduced efficiency of transformation. Alternatively, the presence of the foreign sequence of DNA may lead to modification of the donor DNA. This modification does not extend beyond the intergenotic region because the efficiency of transformation for the *his*B2 marker,which is genetically unlinked to the rifampin resistance marker,is not affected when donor DNA is derived from an intergenotic strain. However,it is not possible to rule out modification of the intergenotic region itself unless intergenotic regions can be constructed with identical nucleotide sequences. The integration of a temperate bacteriophage in the genome of a heterologous strain would appear to provide the most attractive means of determining the extent of modification in the heterologous region.

(b) Efficiency of transfection with heterologous strains lysogenic for Ø105

In order to construct a heterologous strain which was lysogenic for Ø105, it was necessary to infect B. subtilis HSR with Ø105 and select clones which were resistant to Ø105 clz(see Materials and Methods).DNA isolated from these resistant strains was capable of transfection but was restricted in its ability to transform auxotrophic mutants of B. subtilis 168. In order to be assured that the plaques resulted from the release of Ø105 and not some other bacteriophage, the cells were also plated on indicator cells that were lysogenic for Ø105. No plaques were produced on this indicator demonstrating that the cells were only releasing wild type bacteriophage Ø105.The efficiency of transformation and transfection for DNA from B.subtilis HSR(Ø105) and B. subtilis BR151 (Ø105) is shown in Table IV. Although the prophage region of the DNA has the same nucleotide sequence, transfection is markedly reduced when the donor DNA is from the heterologous strain.

The biological activity of prophage DNA from strain HSR was also assayed by the formation of transformants that were resistant to Ø105 clz.The values obtained in this assay were equal to the plaque forming ability of the population transfected with prophage DNA from both BR151 (Ø105) and HSR (Ø105) although the latter was still markedly less efficient (Table V). This assay indicated that the inefficiency of transfection was not due to the inability of the recipient to produce phage after transfection with heterologous DNA since the number of cells which expressed the phage function (resistance to Ø105--presumably by synthesis of a repressor) were equal to the number of plaque forming units.

TABLE IV

EFFICIENCY OF TRANSFECTION AND TRANSFORMATION OF RUB300
WITH PROPHAGE DNA FROM BR151 (Ø105) AND HSR (Ø105)

Donor DNA	Transfectants per 10^8 cells		His^+ transformants per 10^8 cells	Relative efficiency†	
	on 168 Indicator	on HSR Indicator		p.f.u.	His^+
BR151 (Ø105)	1,040	819	18,300	1.0	1.0
HSR (Ø105)	20	68	22	0.02	0.001

† The relative efficiency is the efficiency of transformation or transfection of HSR (Ø105) compared to BR151 (Ø105).

TABLE V

EFFICIENCY OF TRANSFECTION AND TRANSFORMATION OF RUB300 WITH PROPHAGE DNA FROM BR151 (Ø105) AND HSR (Ø105)

Donor DNA	Transfectants p.f.u. per 10^8 cells	Cells resistant to Ø105 clz
BR151 (Ø105)	5,060	6,670
HSR (Ø105)	409	444

(c) Efficiency of transfection and transformation of intergenotic prophage DNA

Selection for strains having an intergenotic prophage region would not be reliable if clones were picked from the center of a plaque that was formed by a transfected cell because there would be no way to separate the original transfectants from the recipient cells which were subsequently infected by the released phage. However, the resistance of transformed cells to Ø105 clz does provide a means for isolating recipient cells having an intergenotic prophage region (a prophage region donated by a heterologous strain). Four intergenotic strains resulting from transformation of BR151 with DNA from HSR (Ø105) and a control strain of BR151 transformed with DNA from BR151 (Ø105) were isolated by this technique. As indicated in Materials and Methods these strains were resistant to Ø105 clz and could produce wild type Ø105. In addition, DNA from these strains could transform *B.subtilis* RUB300 to histidine independence at a high frequency.

The results of transformation and transfection with DNA from these intergenotic strains are given in Table VI. Transformation of RUB300 to histidine prototrophy was as efficient as it was with homologous DNA as expected, since the unlinked marker is in the neighborhood of homologous DNA. The efficiency of transformation to His^+, therefore, has been used as an internal control to determine the efficiency ratio (Wilson & Young, 1972). Transfection with intergenotic prophage DNA, on the other hand, was only 45% as efficient as homologous DNA even though the nucleotide sequence was the same in both strains. The prophage DNA therefore, had the same pattern of efficiency as had been observed with the rifampin resistance and lysine markers (Tables II and III) in the heterologous and intergenotic configurations.

(d) Efficiency of transformation and transfection of a lysogenic recipient

The previous studies with prophage DNA still did not resolve whether the lower efficiency of genetic exchange of the intergenotic donor DNA was due to modification of the intergenotic region or a difference in nucleotide sequence. The possibility remained that heterologous DNA surrounded the prophage region. This small amount of foreign DNA could interfere significantly with integration of the prophage due to inefficient base pairing. To determine if this was the source of inefficient transfection with intergenotic strains, recombination within the prophage region was examined with both heterologous and homologous lysogenic strains. Suppressor mutants of Ø105 were introduced into the two strains, *B. subtilis* HSR and BR151. The two suppressor mutants *sus* 11 and *sus* 19 have been mapped and were found to

TABLE VI

EFFICIENCY OF TRANSFORMATION AND TRANSFECTION OF RUB300 WITH INTERGENOTIC DNA CARRYING PROPHAGE Ø105

Donor DNA Strain	Donor DNA Lysogen	P.F.U. per 10^8 Recipient Cells	His^+ Transformants Per 10^8 Recipient Cells	Efficiency Ratio†
RUB310	BR151(Ø105)	6,400	36,500	1.0
RUB407	BR151(HSR Ø105)	3,060	44,300	0.39
RUB408	BR151(HSR Ø105)	3,760	43,200	0.50
RUB409	BR151(HSR Ø105)	3,290	40,100	0.47
RUB410	BR151(HSR Ø105)	2,900	38,800	0.43

†

$$\text{Efficiency ratio} = \frac{\dfrac{\text{homologous His}^+}{\text{intergenotic His}^+}\ \text{intergenotic transfection (Ø105)}}{\text{homologous transfection (Ø105)}}$$

belong to genes J and C respectively(Armentrout & Rutberg,1970). These genes are separated by approximately 60% of the distance of the genome of Ø105. When the recipient strain BR151 was lysogenic for Ø105 sus 11,wild type Ø105 could be produced by transformation with donor DNA isolated from strains lysogenized with Ø105 sus 19 (Table VII). In control experiments, plaques could

TABLE VII

RECOMBINATION OF BR151 (Ø105 SUS 11) WITH DNA FROM OTHER LYSOGENIC STRAINS

Donor DNA		Wild-type recombinants of Ø105 per 10^8 recipient cells	Efficiency
Strain	Lysogen		
Experiment 1			
RUB311	BR151(Ø105 sus 11)	0	NA†
RUB511	HSR(Ø105 sus 11)	0	NA†
RUB316	BR151(Ø105 sus 19)	1.40×10^4	1.0
RUB516	HSR(Ø105 sus 19)	1.30×10^4	0.93
RUB518	HSR(Ø105 sus 19)	1.40×10^4	1.0
Experiment 2			
RUB316	BR151(Ø105 sus 19)	1.00×10^4	1.0
RUB516	HSR(Ø105 sus 19)	1.51×10^4	1.5
RUB518	HSR(Ø105 sus 19)	1.23×10^4	1.2

† Not applicable

not be produced in the non-permissive host with donor DNA carrying the same suppressor mutation as the recipient.Thus by requir ing recombination within the prophage region for the production of wild type Ø105, it is possible to ascertain the degree of restriction imposed upon transfection by the neighboring heterologous DNA. The results of transfection experiments with prophage DNA (carrying suppressible mutations) from either a heterologous or homologous source clearly demonstrate that the neighborhood of DNA surrounding the prophage region does not affect recombination within the prophage genome itself. DNA isolated from lysogenic strains derived from *B. subtilis* HSR are as efficient agents of transfection as prophage DNA from the homologous background. On the other hand, with similar DNA preparations, when recombination necessarily occurred outside the prophage genome in order to integrate Ø105 (Table IV, line 2), the efficiency of recombination was much lower and was more like that observed for other markers (Table II). If modification of the prophage genome had occurred during passage through the heterologous strains (RUB 511, 516 and 518;Table VII), then recombination should have been equal in strains BR151(Ø105 *sus* 11) and BR151 (compare Table VII and Table IV). The large discrepancy in the efficiency of transfection in these different recipients, therefore, indicates that the efficiency of recombination is not affected by modification but may be determined by some other factor.The most likely limiting factor is the relative degree of homology in the base sequences of the donor and recipient genomes near the region of the intergenote.

4. DISCUSSION

Among the several genospecies in which heterospecific transformation has been studied,it is evident that broad generalizations cannot be made concerning the conditions affecting the efficiency of genetic exchange by heterologous DNA. For example, the lethal consequence of transformation by heterologous DNA documented in the *Haemophilus influenzae* system (Beattie & Setlow, 1970) does not seem to be significant in the *B.subtilis*, pneumococcus, or streptococcus (Biswas & Ravin, 1971) genospecies, although in all of these systems heterologous DNA seems to be absorbed by the cell as efficiently as homologous DNA. In addition, at least in *H. influenzae*, heterologous DNA is integrated as efficiently as homologous DNA (Beattie & Setlow, 1970). These systems do, however, share a common feature in their patterns of restriction for transformation by heterologous DNA (Wilson & Young, 1972; Beattie & Setlow, 1970; Schaeffer, 1958) in that the restriction is partially relieved if the heterologous marker is donated by an intergenote, a strain having a segment of foreign DNA integrated into its genome.

At least two different mechanisms have been proposed which could explain the partial efficiency of the intergenote. One is that the DNA of the donor is modified and subsequently restricted in the recipient cell. Alternatively, the nucleotide sequence of the region itself may become more efficient in its ability to pair with the recipient due to the increased homology of the neighborhood of homologous DNA surrounding the intergenotic marker (Biswas & Ravin, 1971 ; Schaeffer, 1958). Although it is possible that restrictive mechanisms are very important in determining the efficiency of transformation by heterologous DNA (Wilson & Young, 1972) it seems unlikely that they are the major factors which decrease transformation with intergenotic DNA. If classical modification and restriction mechanisms (Boyer, 1971) were operative, one would predict that the DNA would assume the modification pattern of the donor and an intermediate efficiency could not exist unless several genes contributed to the modification of DNA. Of course, it could be argued that the intergenotic region provides either the information for the synthesis of restrictive enzymes or causes the derepression of host-mediated re strictive enzymes. In some cases, then, establishment of the intergenote would not include all of these foreign regions, thus producing intermediate levels of restriction. However, if this were the explanation, it would be difficult to understand why markers that are genetically unlinked from the intergenotic region are also not subject to the same level of restriction. From Table III, it is clear that unlinked markers are not restricted in their efficiency of transformation. Thus, the only explanation which invokes the mechanism of modification that can not be ruled out is modification of only the intergenotic region by more than one modifying gene that is either present in the heterologous DNA or is derepressed by the presence of heterologous DNA in the intergenotic strain.

The partial efficiency of the intergenote can also be explained on the basis of the nucleotide sequence of the region. Obviously, regions of the genomes which share little homology can not recombine because pairing is markedly restricted. However, if the intergenote were surrounded by homologous DNA, this neighborhood could enhance genetic exchange. This explanation, although the more attractive of the two, could not be resolved from the first until an intergenote could be constructed which had the same nucleotide sequence in both the heterologous and homologous donors. By using temperate bacteriophage Ø105, we have been able to satisfy these requirements. In Figure 2 we have diagrammed our interpretation of the genetic constitution of these strains based on the genetic data obtained with these donors. When the bacterio phage is integrated in the strain which is homologous to the recipient, the neighborhood of the prophage region is optimal for

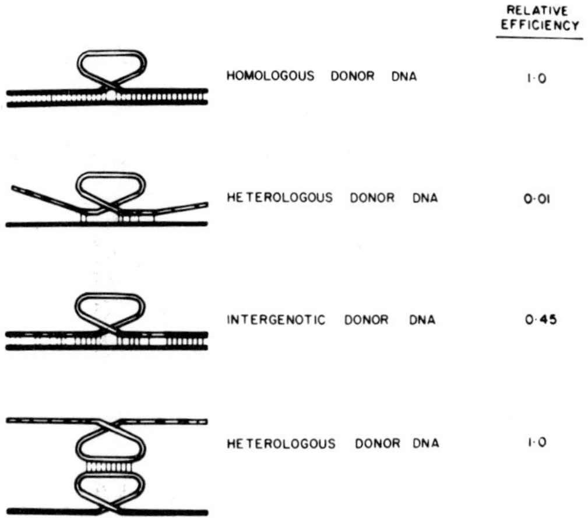

Fig. 2 - The relative efficiencies of transfection for DNA from lysogenic strains. Double-stranded DNA is represented with two heavy bars. Regions of nucleotide sequences that are identical to sequences in B.subtilis 168 are solid black. Light vertical lines indicate proper base pairing between the two strands of DNA. The loop designates the prophage genome.

genetic exchange to occur. However, when the prophage is integrated in the heterologous strain, there are no major regions of homology and genetic exchange is severely restricted. An intermediate case exists when the intergenote is constructed. Here only small regions of heterologous DNA surround the prophage and these regions only slightly impair efficient exchange because sufficient homology exists outside the regions in the neighborhood. Thus far, these strains have displayed the same pattern of restriction which is common to transformation for auxotrophic markers in homologous, heterologous, and intergenotic donors. In addition, with these constructions it is possible to examine recombination within the intergenotic region itself. By using a lysogenic recipient, a region of homology is therefore created between the lysogenic strains regardless of the rest of the genome. The higher levels of recombination between the suppressible mutations within the genome of bacteriophage Ø105 therefore supports the contention that the neighborhood is the primary determinant of recombination.

These studies have also expanded the possibilities for genetic exchange among heterologous organisms. It seems reasonable to assume that genetic exchange between species could be increased if regions of homologous nucleotide sequences could be introduced in both strains. Conceivably, these genetic modifications need not be limited to crosses between bacterial species, but may also include interaction with eucaryotic systems. For example two reports have presented biological evidence for genetic exchange between bacteria and other organisms. Ledoux, Huart, and Jacobs (1971) showed that a tryptophan requiring mutant of *Arabidopsis thaliana*(L.)Heynt could grow in the absence of tryptophan after treating the seeds with DNA from wild type *B. subtilis* cells. Merril, Geier, and Petricciani(1971)reported α-D-galactose-1-phosphate uridyl transferase activity in galactosemic human fibroblasts after treating the cells with DNA from defective lambda bacteriophage. In addition to these studies, *B. subtilis* has been shown to be transformed to methionine independence with DNA from the complex hyphae producing bacteria, *Streptomyces coelicolor* (Mergeay, Charles, Martin, Remy & Ledoux, 1971).

In all of these investigations, the frequency of recombination was very low; however, if techniques were developed to incorporate DNA in or near a prophage region, it might be possible to incorporate segments of foreign genomes at significant levels as has been done here for the *B. subtilis* system.

ACKNOWLEDGEMENTS

This work was supported by grant VC-27H from the American Cancer Society and grant GB-24379 from the National Science Foundation. We wish to thank the American Society for Microbiology for permission to reproduce Figure 1 and to summarize data from a previous publication (Wilson & Young, 1972).

5. REFERENCES

ARMENTROUT, R. W. & RUTBERG, L. (1970). J. Virol. 6:760.

BEATTIE, K. L. & SETLOW, J. K. (1970). J. Bacteriol. 104:390.

BISWAS, G. D. & RAVIN, A. W. (1971). Mol. Gen. Genetics 110:1.

BOYER, H. W. (1971). Ann. Rev. Microbiol. 25:153.

CHILTON, M-D. & MCCARTHY, B. J. (1969). Genetics 62:697.

DUBNAU, D., SMITH, I., MORELL, P. & MARMUR, J. (1965). Proc. Nat. Acad. Sci., Wash. 45:491.

GEORGOPOULOS, C. P. (1969). J. Bacteriol. 97:1397.

GROMKOVA, R. & GOODGAL, S. H. (1972). J. Bacteriol. 109:987.

KELLY, T. J. & SMITH, H. O. (1970). J. Mol. Biol. 51:393.

LEDOUX, L., HUART, R. & JACOBS, M. (1971). In Informative Molecules in Biological Systems, ed. by L. Ledoux, p. 159, Amsterdam: North Holland Pub. Co.

LOVETT, P. S. & YOUNG, F. E. (1969). J. Bacteriol. 100:658.

MERGEAY, M., CHARLES, P., MARTIN, R., REMY, J. & LEDOUX, L. (1971). In Informative Molecules in Biological Systems, ed. by L. Ledoux, p. 51, Amsterdam: North Holland Pub. Co.

MERRIL, C. R., GEIER, M. R. & PETRICCIANI, J. C. (1971). Nature (London), 233:398.

MIDDLETON, J. H., MARSHALL, E. H. & HUTCHISON, C. A. III. (1972) J. Virol. 10:42.

NESTER, E. W., SCHAEFFER, M. & LEDERBERG, J. (1963). Genetics 48: 529.

RAVIN, A. W. (1963). Amer. Natur. 97:307.

REILLY, B. E. & SPIZIZEN, J. (1965). J. Bacteriol. 89:782.

RUTBERG, L. & ARMENTROUT, R. W. (1970). J. Virol. 6:768.

SCHAEFFER, P. (1958). Symp. Soc. Exp. Biol. 12:60.

SMITH, H. O. & WILCOX, K. W. (1970). J. Mol. Biol. 51:379.

SPIZIZEN, J. (1958). Proc. Nat. Acad. Sci., Wash. 44:1072.

WILSON, G. A. & YOUNG, F. E. (1972). J. Bacteriol. 111:705.

YASBIN, R. E. & YOUNG, F. E. (1972). Biochem. Biophys. Res. Commun. 47:365.

YOUNG, F. E. & WILSON, G. A. (1972). In Spores V, ed. by H. O. Halvorson, R. Hanson & L. L. Campbell, p. 77. Washington, D. C.:American Society for Microbiology.

EXTENT OF DIPLOIDY IN STABLE AND UNSTABLE MEROZYGOTES FROM *BACILLUS SUBTILIS* STRAINS BEARING THE *trpE26* MARKER

C. AUDIT

C. ANAGNOSTOPOULOS

Centre de Génétique Moléculaire, C.N.R.S.,
91190, GIF-SUR-YVETTE, France

ABSTRACT

In the merozygotes appearing after transformation of *B. subtilis* strains that carry the *trpE26* marker, the diploid condition is not limited to genetic loci proximal to this marker. It was shown that in the unstable merozygotes many regions of the greater part of the chromosome (about 80% its total length) may be involved in the diploid state.Only loci of the part closer to the origin of replication were never found to be duplicated. In the majority of the stable merozygotes the diploid region is confined to the right side of *trpE26*. The duplicated areas are often very large.Genetic loci situated between two heterozygous regions were found to be in an homodiploid state thus demonstrating continuity of the diploid area. Results are compatible with the hypothesis that in the *trpE26*$^{+}$/*trpE26* merozygotes a long chromosomal segment is duplicated. Whether the second copy of this segment is integrated into the chromosome or is carried as an episome is not yet understood.

1. INTRODUCTION

In a previous paper (Audit and Anagnostopoulos, 1972) we reported evidence on the production of merodiploid clones in

Bacillus subtilis. These organisms arise after transformation or PBS-1-mediated transduction of strains bearing the trpE26 marker. TrpE26 is a mutation in the anthranilate synthetase locus which is part of the "aromatic segment" of the chromosome. Tryptophan independent transformants or transductants for this marker are usually unstable and segregate tryptophan requiring cells. Their growth is inhibited by L-lysine. However some of them (2 to 5% among the transformants and about 50% among the transductants) become stabilized in the diploid state and are, in their great majority, resistant to lysine. There is an important difference between this system of merodiploid formation and those reported for other organisms (Bernheimer, Vermundsen and Austrian, 1967; Curtiss, 1964; Hill, Schiffer and Berg, 1969; Ravin and Takahashi, 1970; Hotchkiss, Abe and Lane, 1971). In the B. subtilis heterogenotic strains the diploid condition is not limited to the chromosomal region in the vicinity of the specific marker. Loci very distant from the "aromatic segment" are often found to be diploid. Experiments reported in this paper deal with this question of the extent of merodiploidy in unstable as well as in stable $trpE26^+$/trpE26 heterogenotes.

2. RESULTS AND DISCUSSION

(a) Merodiploidy in chromosomal regions unlinked to trpE26.

In order to ascertain whether parts of the genome other than loci in the "aromatic segment" could become merodiploid we transformed several double mutants carrying trpE26 (general genotype trpE26 A; where A stands for an auxotrophic marker unlinked to trpE26) by wild type DNA and examined the progeny of single trpE26 A^+ and double $trpE26^+$ A^+ transformants. For the first we observed, after purification, that, whatever the A marker, they were always pure clones, containing only A^+ cells. It appears therefore that in strains carrying trpE26, if selection is not made for Trp^+ (tryptophan independence) only haploid transformants are obtained. These results seem not to support the hypothesis that the total cell population in a culture of a trpE26 strain is diploid for the whole or part of the genome.

The results with the prototrophic double transformants (which were unstable), were entirely different (Table I). Among eighteen Trp^+ Ilv^+ clones examined, as many as ten segregated not only Trp^- but also Ilv^- cells (Ilv^-: requirement for isoleucine and valine). These are double merozygotes $trpE26^+$/trpE26 $ilvC1^+$/ilvC1. Similar results were obtained with the thr-5 marker (Thr: requirement for threonine): ten among eighteen Trp^+ Thr^+ transformants turned out to be double merozygotes $trpE26^+$/trpE26 $thr\text{-}5^+$/thr-5. However Pur^- or Cys^- (Pur: requirement for purine bases; Cys: requirement for cysteine) segregants were never found among the progeny of the double transformants from the three last crosses of Table I (with recipients bearing markers purA16, cysA14 or purB6). This suggests that the regions of the genome corresponding to these loci are never (or very seldom) diploid. The fact that these are genes located near the origin of the chromosome (Yoshikawa and Sueoka, 1963; Dubnau et al, 1967) is probably significant.

In the case of the double merozygotes all combinations of the two unlinked markers (Trp^+ A^+, Trp^+ A^-, Trp^- A^+ and Trp^- A^- phenotypes) were often found among the progeny of a single transformed cell (Table I). This segregation pattern suggests that recombination events may occur in the diploidized region.

Evidence for a merodiploid condition was also sought

TABLE I

SEGREGATION OF UNSTABLE DOUBLE TRANSFORMANTS FOR TWO UNLINKED MARKERS[+]

Genotype of recipient strain trpE26 A	Transformants				
	Ratio: Clones segregating A^- cells / No of prototrophs examined	Frequencies of the different classes of cells[‡] (phenotype)			
A		$Trp^+ A^+$	$Trp^+ A^-$	$Trp^- A^+$	$Trp^- A^-$
ilvC1	10/18	0.27 - 0.60	0.0 - 0.18	0.10 - 0.63	0.0 - 0.33
thr-5	10/18	0.23 - 0.76	0.0 - 0.015	0.21 - 0.63	0.0 - 0.14
purB6	0/10	0.08 - 0.50	0	0.50 - 0.92	0
cysA14	0/13	0.24 - 0.67	0	0.33 - 0.76	0
purA16	0/11	0.04 - 0.25	0	0.75 - 0.96	0

\+ The recipient strains were transformed by wild type (WT) DNA at saturating concentration (> 1.0 μg/ml). Unstable prototrophic transformants were tested for segregation after several purifications.

‡ Figures represent the range of the frequencies of the various classes of bacteria found in the clones tested.

by testing the DNA of the prototrophic double transformants for the presence of auxotrophic markers derived from either the donor or the recipient strain. This method is moreover the only one available for the study of the stable clones where the other approach (study of the segregation pattern) is not feasible.

The results of Table II illustrate the use of this method. After transformation of a strain the genotype of which was trpE26 ilvC1 by DNA from a strain carrying the leu-2 marker (Leu: requirement for leucine; leu-2 is linked to ilvC locus) eighteen prototrophic transformants were purified. Their DNA was extracted and used as donor in two different transformation experiments. In the first experiment the genotype of the recipient was leu-2. The occurrence of Leu^+ Ilv^- transformants reveals that the donor prototrophic strain possesses the ilvC1 marker. In the second control experiment an ilvC1 recipient was used. This transformation makes it possible to detect the leu-2 marker, the occurrence of Ilv^+ Leu^- transformants revealing the presence of this marker in the donor DNA. In this last experiment among eighteen prototrophs examined (all of which segregated Trp^- cells) seven did not possess either leu-2 or ilvC1 and were considered to be haploid in the ilvC-leu region. However the possibility that these clones might be homodiploid $ilvC1^+$ $leu\text{-}2^+$/ $ilvC1^+$ $leu\text{-}2^+$ cannot be excluded. Five clones were found to possess ilvC1 alone and were probably double merozygotes $trpE26^+$/ trpE26 $ilvC1^+$ $leu\text{-}2^+$/ilvC1 $leu\text{-}2^+$. The remaining six clones possessed both ilvC1 and leu-2. Their genotype most probably is $trpE26^+$/trpE26 $ilvC1^+$ leu-2/ilvC1 $leu\text{-}2^+$.

Among the eighteen unstable prototrophs from the above initial transformation (Table II) eleven were therefore merodiploid in the ilvC - leu region for at least the ilvC1 lo

TABLE II

OCCURRENCE OF leu-2 AND ilvC1 MARKERS IN THE DNA OF PHENOTYPICALLY PROTOTROPHIC UNSTABLE MERODIPLOIDS+

Recipient (relevant genotype)	Ratio of transformants	No. of donor clone (unstable prototrophic transformants from the initial transformation)										
		1	2	3	4	5	6	7	8	9	10	11
ilvC1	$\frac{Ilv^+ Leu^-}{Ilv^+}$	0	0.52	0.53	0	0.40	0.50	0.22	0.62	0	0	0
leu-2	$\frac{Leu^+ Ilv^-}{Leu^+}$	0.09	0.37	0.33	0.06	0.28	0.26	0.34	0.58	0.28	0.16	0.20

\+ Initial transformation: recipient (genotype) trpE26 ilvC1
donor (genotype) leu-2.

Eighteen clones of wild type phenotype from the initial transformation were purified and their DNA was used as donor in two different control transformation experiments involving recipient strains carrying ilvC1 and leu-2.
The ratios of the transformants $\frac{Ilv^+ Leu^-}{Ilv^+}$ and $\frac{Leu^+ Ilv^-}{Leu^+}$ were measured. For seven of the donor clones examined both these ratios were nil (results not presented).

TABLE III

SIMULTANEOUS OCCURRENCE OF hisH2 AND ilvA2 MARKERS IN THE DNA OF PHENOTYPICALLY PROTOTROPHIC STABLE MERODIPLOIDS^{+}

Recipient (genotype)	No. of donor clone (stable prototrophic transformants of the initial transformation)	Ratio of transformant classes	
		$Trp^{+}\ His^{-}$ / Trp^{+}	$Met^{+}\ Ile^{-}$ / Met^{+}
trpC2 metB3	1	0.39	0.09
	2	0.24	0

$^{+}$ Initial transformation: Recipient trpE26 ilvA2
donor hisH2.

Two stable prototrophic transformants, obtained in the initial transformation, were used as donors in a control transformation experiment with a recipient strain carrying trpC2 and metB3. (The metB3 and ilvA2 markers are linked).

cus. On the contrary, when we examined stable prototrophs derived from the same initial transformation no indication of heterozygosity at this region was revealed. These must be either haploid $ilvC1^{+}$ $leu\text{-}2^{+}$ or homodiploid.

However, stable heterogenotes for a region other than the "aromatic segment" have been isolated. These were transformants for the ilvA locus, which lies on the right side of trpE26 (Table III). A trpE26 ilvA2 recipient was transformed by DNA from a strain bearing the hisH2 marker, linked to trpE26. Two stable prototrophic transformants were analyzed. When used as donors both transmitted the hisH2 allele. One of them also transmitted the ilvA2 marker and was thus shown to be a merozygote in at least two regions: $trpE26^{+}$ hisH2/trpE26 $hisH2^{+}$ $ilvA2^{+}$/ilvA2.

Similar experiments were carried out with strains bearing *trpE26* and a variety of other markers (*pheA1* and *aroB2* on the left side of the tryptophan cluster; *thyB1* and *metB3* on the right side of it). The detailed results of these crosses will be reported elsewhere (*manuscript in preparation*). They were very similar to those reported above. All these results suggest that any region of the genome from *thr-5* to *ilvA* can exist in a diploid state. This interval corresponds to about 80% of the known length of the chromosome. The remaining 20%, close to the origin of replication, do not seem ever to become merodiploid. The stable merozygotes seem (at least in their majority) to be diploid only for the regions on the right side of the *trpE26* marker.

(b) Continuity of the diploid state in a large area of the chromosome.

As was pointed out above, in order to isolate clones heterogenotic for a locus different than *trpE* it appears essential to select simultaneously transformants for *trpE26*$^{+}$ and for the marker of the other locus.

It was later found that strains already merodiploid in the "tryptophan region" also yield clones diploid in other regions, in consecutive transformation experiments. Stable as well as unstable diploids are obtained and the same restrictions as regards the position of the second marker on the chromosome apply in this case too as in that of simultaneous selection (*manuscript in preparation*).

These observations raised the question as to whether in *trpE26*$^{+}$/*trpE26* heterogenotes the diploid state might not involve a very large part of the chromosome. In that case subse-

TABLE IV

HETEROZYGOTIC CONDITION AFTER REVERSION FOR A LOCUS SITUATED BETWEEN TWO DIPLOIDIZED REGIONS+

Donor strain (phenotype)	Recipient strain (genotype)	Ratio of transformants ‡
Revertant clone	leu-2	$\frac{Leu^+ Ilv^-}{Leu^+}$: 0.26
Trp^+ Phe^+ Ilv^+	nic-38	$\frac{Nic^+ Phe^-}{Nic^+}$: 0.24

+ A triple auxotroph the genotype of which was trpE26 pheA1 ilvC1 was initially transformed by wild type DNA at saturating concentration. Double transformants Trp^+ Phe^- Ilv^+ were selected and purified. Spontaneous revertants for the pheA1 marker were then sought from these clones. One such revertant was finally used as donor in the control transformation experiments.

‡ The cotransfer frequencies measured in normal crosses are the following: for ilvC1 and leu-2: 0.40 (Barat, Anagnostopoulos and Schneider, 1966); for pheA1 and nic-38: 0.62 (Hoch and Anagnostopoulos, 1970).

quent transformations do not create diploidy in other regions but just reveal (by transforming homozygous cells to heterozygous) a preexistent homodiploidy.

The results shown in Table IV, which demonstrate the continuity of the diploid condition between two distant heterozygous regions, are in good agreement with the above hypothesis.

In these experiments a trpE26 pheA1 ilvC1 recipient was transformed by WT DNA and a double transformant Trp^+ Phe^- Ilv^+, segretating Trp^- and Ilv^- cells, was selected. This strain was heterogenotic for the two transformed loci (trpE and ilvC).

TABLE V

HETEROZYGOTIC CONDITION AFTER REVERSION FOR THE ilvC LOCUS IN AN UNSTABLE MERODIPLOID+

Donor strain (phenotype)	Recipient strain (relevant genotype)	Ratio of transformants Leu^+ Ilv^-/Leu^+
Revertant clone No.1	leu-2‡	0.25
Revertant clone No.2	leu-2	0.14

+ A strain the genotype of which was trpE26 ilvC1 was initially transformed by wild type DNA. Trp^+ Ilv^- transformants were selected and purified. From these clones spontaneous revertants for the ilvC1 marker were obtained. After purification these Trp^+ Ilv^+ clones were used as donors in the control transformation experiments.

‡ For the linkage of leu-2 to ilvC1 see Table IV.

Since pheA1 reverts spontaneously (at low frequency) a Phe^+ revertant was isolated from the above clone. After extracting its DNA and using it as donor in transformation experiments with suitable recipients it was found that this prototrophic clone possessed not only the ilvC1 marker as expected but also the pheA1 marker. The presence in the same clone of the $pheA^+$ allele was ascertained after transforming with its DNA a strain bearing pheA1. The Phe^+ transformants obtained in this last experiment did not any longer possess this auxotrophic marker. The phenylalanine independence displayed by the revertant did not therefore result from an external suppressor mutation but from a mutation at the original site, or at a site very close to it, in one of the pheA1 copies of the heterogenote. The above results have established the fact that the genotype of the double merozygote, Trp^+ Phe^- Ilv^+, which gave rise to the revertant,

TABLE VI

HETEROZYGOTIC CONDITION AFTER REVERSION FOR THE ilvA LOCUS IN A STABLE MERODIPLOID+

Donor strain (pehnotype)	Recipient strain (relevant genotype)	Ratio of transformants $Met^+ Ile^-/Met^+$
Revertant clone No.1	metB3‡	0.10
Revertant clone No.2	metB3	0.05

\+ A recipient strain carrying trpE26 and ilvA2 (ilvA: requirement for isoleucine) was transformed by WT (wild type) DNA. $Trp^+ Ile^-$ stable transformed clones were selected. After purification of a few of these spontaneous isoleucine independent revertants ($Trp^+ Ile^+$) were sought. Two of them have been used as donors in the control transformation experiments.

‡ The cotransfer frequency of metB3 and ilvA2 in normal repulsion crosses is 0.11.

was: $\frac{trpE26^+ \; pheA1 \; ilvC1^+}{trpE26 \; pheA1 \; ilvC1}$. This clone was indeed homodiploid for the pheA locus, which lies between the two heterogenotic regions on the chromosome map.

By the same method (i.e. selection of prototrophic revertants) we were also able to demonstrate the homodiploid condition for the ilvC locus in an unstable $trpE26^+/trpE26$ single transformant (Table V).

Studies on the reversion of markers in stable merozygotes gave, as expected, quite different results. Prototrophic revertants for those markers located on the left side of trpE26 were never found to be heterogenotic in the regions considered. On the other hand two Ile^+ clones which arose spontaneously in a stable $Trp^+ Ile^-$ transformant were shown to possess both the ilvA2 marker (Table VI) and also (by suitable crosses not presented here) the $ilvA^+$ allele. They were therefore $ilvA2^+$/ilvA2 heterogenotes and the Trp^+ Ile^- transformant from which they were derived must have been homodiploid for the ilvA locus, which is situated on the right side of trpE26 and outside the transformed area.

3. CONCLUSIONS

The results reported in this paper are in good agreement with the hypothesis that the $trpE26^+$/trpE26 merozygotes carry a duplication of a long continuous chromosomal segment. In the case of the unstable clones the duplication might often involve the greater part (up to 80%) of the bacterial genome. As regards the stable merozygotes, in the cases analyzed above, the duplication was found to concern only chromosome areas located on the right side of the trpE26 marker.Whether both copies of the duplicated segment are integrated into the chromosome or one is carried as an episome cannot yet be ascertained. However, preliminary data (mainly from PBS1 mediated transduction experiments) suggest that in stable merozygotes the physical structure of the diploidized region may be that of a tandem duplication. The way the merozygotes are originally formed and whether their formation is related or not to transformation or transduction at the trpE26 site are at present unknown.

4. REFERENCES

AUDIT, C. & ANAGNOSTOPOULOS, C. (1972). *In* L.L. Campbell (ed.) Spores V p. 117, Washington D.C. American Society for Microbiology.

BARAT, M., ANAGNOSTOPOULOS, C. & SCHNEIDER, A.-M. (1965). J. Bact. 90:357.

BERNHEIMER, H.P., WERMUNDSEN, I.E. & AUSTRIAN, R. (1967). J. Bact. 93:320.

CURTISS, R. (1964). Genetics, 50:679.

DUBNAU, D., GOLDTHWAITE, C., SMITH, I. & MARMUR, J. (1967). J. Mol. Biol. 27:163.

HILL, G.W., SCHIFFER, D. & BERG, P. (1969). J. Bact. 99:274.

HOCH, J.A. & ANAGNOSTOPOULOS, C. (1970). J. Bact. 103:295.

HOTCHKISS, R.D., ABE, M. & LANE, D. (1971). *In* L. Ledoux (ed.) Informative Molecules in Biological Systems p. 21, Amsterdam: North Holland Publishing Co.

RAVIN, A.W. & TAKAHASHI, E.A. (1970). J. Bact. 101:38.

YOSHIKAWA, H. & SUEOKA, N. (1963). Proc. Nat. Acad. Sci., Wash. 49:559.

MARKER EFFECTS IN STREPTOCOCCAL TRANSFORMATION

J. G. TIRABY

M. S. FOX

Department of Biology
Massachusetts Institute of Technology
Cambridge, Mass. 02139

ABSTRACT

All of the more than 80 streptococcal markers that we have examined are transferred by transformation with equal efficiencies. These include markers of spontaneous origin as well as some derived by nitrous acid mutagenesis.

Marker effects are manifested in heterospecific transformation of streptococcus with pneumococcal DNA. These effects show a locus dependence and may depend on the local degree of polynucleotide sequence homology between pneumococcus and streptococcus. A number of pneumococcal markers that exhibit characteristic integration efficiencies in discriminating strains of pneumococcus, and whose integration efficiencies are uniformly high in non-discriminating strains, were examined in heterospecific transformation. There is no correlation between the behavior of these markers in the two systems.

There is no evidence, in Group H streptococcus, of a discriminating system analogous to that observed in pneumococcus.

1. INTRODUCTION

For a given quantity of DNA taken up in pneumococcal transformation, individual markers exhibit characteristic transforming activities. Since the yield of transformants for a given marker is proportional to the quantity of DNA taken up(Fox,1957), the presence of a standard marker in a DNA preparation facilitates the measurement of the relative efficiency with which any other marker can be introduced by transformation.

Some markers are demonstrably multi-site(perhaps deletions) and their characteristic efficiencies vary inversely with the extent of the multi-site mutation(Lacks, 1966).An unexpected class of single site markers are those whose efficiency is dependent on the pneumococcal strain chosen as a recipient.

Several pneumococcal strains, widely used in the study of transformation, exhibit striking differences in the efficiencies with which many markers representing point mutations are successfully integrated (Ephrussi-Taylor, Sicard and Kamen,1965; Lacks, 1966; Gray and Ephrussi, 1967; Morse and Lerman, 1969; Sirotnak and Hachtel, 1969;Friedman and Ravin, 1972). In strains which manifest this discrimination, markers in the class corresponding to the highest efficiencies are transmitted 15 to 20 times more frequently than those in the low efficiency classes. Other strains of pneumococcus, as well as several mutants that have been isolated,have lost their capacity to discriminate among these classes of markers (Lacks, 1970;Tiraby and Sicard, in preparation).Transforming DNA is integrated by a single strand displacement mechanism,producing a heterozygous heteroduplex region in the genome of the recipient bacterium(Guerrini and Fox,1968). It seems likely that the discriminating strains carry the capacity to recognize and destroy certain transformants by virtue of the base pair mismatches that they contain within the heterozygous region and that this capacity can be lost by mutation.

Though there is little evidence for marker discrimination in the B. subtilis or the H. influenzae transformation system, the possibility that such discrimination might exist in the Group H streptococci seemed worthy of investigation.

The examination of the integration efficiencies of markers representing independently arising mutations for a number of different loci has yielded no evidence of marker discrimination in either the Challis or Wicky strains of Streptococcus. When pneumococcal DNA carrying characterized markers is used to transform streptococcus, it becomes evident that (a) variable marker efficiencies are manifested; (b) this variability is not related to the discrimination phenomenon exhibited by the discriminating pneumococcal strains,and (c) the efficiencies with which markers are transferred exhibit a locus dependence.

2. MATERIALS AND METHODS

(a) Organisms

Strains Challis and Wicky of Group H Streptococcus,and strain R6, a discriminating strain of type III pneumococcus (Diplococcus pneumoniae), were used as drug sensitive recipients. Streptococcus strains were obtained from Jon Ranhand. Strain Wicky, transformable only with the addition of competence factor (CF)was an erythromycin resistant mutant (resistant to 1.5 μg of erythromycin per ml) (Ranhand, Leonard and Cole, 1971). Human blood too old to be usable for transfusion was obtained from a local blood bank.

Donor DNA's were prepared from derivatives of R6 and Wicky, the latter carrying the pneumococcal marker Sm-41 and one or several markers of pneumococcal or streptococcal origin which will be described in the text.

(b) Growth and maintenance of organisms

Challis and Wicky cells were grown and maintained in a medium (BHI-HS) consisting of Brain Heart Infusion (Difco) supplemented with 2.5% (v/v)heat inactivated (56 to 60°C for 30 min) horse serum (Microbiological Associates) (Ranhand, Leonard and Cole, 1971).

R6 cells were grown in M medium consisting of 10 g trypticase(BBL), 5 g neopeptone(Difco), 1 g yeast extract(Difco), 5 g NaCl, 1.25 g Trizma Base(Sigma)per liter of distilled water. This medium was sterilized by autoclaving; sterile glucose was added to a concentration of 0.2% before use.

Wicky and Challis cells as well as R6 cells were plated in M medium supplemented with 1.25% agar and 2%(v/v)whole blood. Colonies of streptococcus and pneumococcus were counted after 25-30 hours and 20 hours incubation respectively. Streptococcus cells were stored at -30°C in BHI-HS without glycerol. Pneumococcus cells were stored at -70°C in M medium containing 10% sterile glycerol.

(c) Preparation of DNA

Streptococcal DNA : Ten ml of BHI-HS were inoculated with 0.5 ml of a frozen (-30°C) stationary culture of one of the Wicky derivatives.After 5-6 hours growth the cells were harvested by centrifugation, washed once with 5 ml of 0.1M Tris, 0.01M EDTA, resuspended in 1 ml of 0.1M Tris,0.01M EDTA, 1 mg/ml lysozyme, and incubated for 20 min at 37°C. Sarkosyl was added to a concentration of 0.1% and the suspension was transferred to a freezer at -70°C. Thawing at room temperature after overnight storage in the freezer resulted in lysis of the cells. The DNA was precipitated with 2 ml of 95% alcohol, the fibers were withdrawn and dissolved in 0.03M sodium chloride, 0.03M sodium citrate.

From the tables in the text it can be seen that all DNA's are not equally active. The correlation observed between the decrease of the transforming activity and the increasing time separating the lysis of the cells and the addition of alcohol suggests that nucleases might be responsible for low activities shown by some DNA preparations.

Pneumococcal DNA : Ten ml of M medium were inoculated with 0.5ml of a frozen (-70° C) stationary culture of one of the derivatives of strain R6. After 5-6 hours growth the cells were harvested by centrifugation and the pellet resuspended in 1 ml of 0.1M Tris, 0.01M EDTA, 0.1% sodium desoxycholate.After 10 min at 37°C to allow the lysis of the cells,the DNA was precipitated with 2 ml alcohol. The fibers were withdrawn and dissolved in .03M sodium chloride, .003M sodium citrate. The pneumococcal DNA preparations were uniform with regard to their transforming activities.

(d) Transformation

Transformation of Challis and Wicky were performed in a manner similar to that described by Ranhand, Leonard and Cole (1971).

Challis: Thawed Challis cells(0.5 ml)were diluted with BHI-HS (10 ml) and incubated overnight at 37°C.One millimeter of this culture was added to 20 ml of prewarmed BHI-HS medium. The cells were competent after 2 hours incubation at 37°C. Competent cells were exposed to a saturating amount of DNA, incubated for 30 min,diluted 5-fold with BHI-HS and incubated for an additional 3 hrs to allow phenotypic expression of transformants.Then cells were plated after appropriate dilution in M medium supplemented with whole blood and the selective agent for scoring transformants.

Wicky: One milliliter from an overnight Wicky culture was diluted with 16 ml of prewarmed BHI-HS. After 50 min incubation, 4 ml of crude competence factor were added to the growing culture. The cells were fully competent after an additional 50 min incubation. The subsequent steps were the same as those for Challis. When the Wicky strains requiring thymidine were used as recipients,thymidine was added to BHI-HS at the concentration of 50 µg/ml. In our hands, the yield of transformants is always significantly higher in Wicky than in Challis.

R6: R6 cells were made competent by the procedure of Gurney and Fox (1968).

(e) Preparation of Challis competence factor

Five milliliters from an overnight Challis culture were diluted with 90ml of prewarmed BHI-HS and 10ml of 10% neopeptone (Difco).After 3 hours incubation without shaking,the culture was centrifuged (10,000 rpm/min in a Sorvall centrifuge,5°C, 15 min), the supernatant was distributed in several Erlenmeyer flasks and immediately stored in a freezer at -30°C. Thawed supernatant was used to induce competence in Wicky cells.

(f) Isolation of Wicky strains requiring thymidine

Strain W thy$^-$A2 was isolated from Wicky after transformation by DNA from a pneumococcal strain R6 thy$^-$ carrying the marker thy$^-$A2 (Friedman and Ravin, 1972). Strain W thy$^-$1 was isolated as a spontaneous mutant of Wicky.

The screening procedure for transformants or the mutant W thy$^-$1 was the following: various dilutions of a Wicky culture grown for 5 hours after transformation by DNA and of a Wicky culture grown in a similar manner without having been exposed to DNA were plated in M medium supplemented with blood,100 µg/ml of thymidine and 15 µg/ml of aminopterin. Resistant colonies were picked and tested for ability to grow in the absence of thymidine. In the transformed culture 90% of the aminopterin resistant colonies were thy$^-$. In the control culture only 4 thy$^-$ colonies were found, corresponding to a frequency of about 100 times fewer than in the transformed culture.

Strains W thy$^-$A2 and W thy$^-$1 both require 50 µg/ml thymidine. As in pneumococcus thymine cannot supplement the nutritional deficiency.

(g) Antibiotics

Streptomycin sulfate USP was purchased from Eli Lilly and Co., Indianapolis.Aminopterin was purchased from Nutritional Biochemical Corp., Cleveland. Streptolydigin was a gift of the Upjohn Co.,Michigan.Fusidic acid was a gift of Leo Pharmaceutical

Products, Ballerup, Denmark. Rifampin was a gift of the Ciba Pharmaceutical Co., New Jersey. Kanamycin was a gift of Bristol Laboratories, Syracuse, New York.

3. RESULTS

(a) Homospecific transformation

According to their integration efficiencies, spontaneous mutants of Diplococcus pneumoniae were distributed with almost equal numbers in the high and low efficiency classes (Ephrussi-Taylor, Sicard and Kamen, 1965; Lacks, 1966; Tiraby and Sicard, in preparation). With this observation in mind, the integration efficiencies of a number of markers of spontaneous origin were measured. The integration efficiency of a given marker is the ratio of the number of transformants for this marker to the number of transformants observed for a reference marker. As a reference marker, we used either a marker conferring resistance to streptomycin Sm-r 41 or a marker conferring resistance to rifampin Rif-r 1 (Ranhand and Cole, 1971). The loci investigated were those determining resistance to fusidic acid (fus-r), rifampin (Rif-r) and streptolydigin (stg-r). Mutations resulting in resistance to at least 75 μg/ml of fusidic acid are closely linked to the reference marker Sm-r 41. These fus-r markers might be mutants of the G translocation factor (Kuwano, Schlessinger, Rinaldi, Felicetti and Tocchini-Valentini, 1971). Mutations conferring resistance to at least 20 μg/ml of rifampin are closely linked to sites conferring resistance to at least 15 μg/ml of streptolydigin. Rif-r and stg-r markers might be mutants of RNA polymerase (Wehrli and Matthys, 1971).

In order to isolate independently arising spontaneous mutants, an overnight culture of strain Wicky carrying markers Sm-r 41 or Rif-r 1 was used to inoculate 50 ml of prewarmed BHI-HS medium with about 10^3 colony-forming units/ml and distributed into 50 small tubes.After 9-10 hours incubation at 37°C,the contents of each tube was poured into a Petri dish and embedded with M medium agar containing one of the 3 antibiotics and incubated overnight at 37°.One of the largest colonies from each plate was used as an inoculum for a small culture from which DNA was isolated and used for transformation.The transforming activities of the reference marker and the marker characteristic of each of the mutants isolated were determined.

The integration efficiencies for markers from 10 independent fusidic acid resistant mutants using both Wicky and Challis as recipient strains are shown in Table I. Within the limits of certainty of the measurement,about ±15%, all of these markers are integrated with equal efficiencies and the two streptococcal strains are indistinguishable from each other with regard to their capacity to integrate these markers.The analysis of 30 independently isolated spontaneous rifampin resistant mutants(Table II)as well as 30 streptolydigin resistant mutants yielded similar results.

As a limited representation of auxotrophic markers, 2 thymidine requiring mutants of Wicky were isolated.These mutants were made competent, transformed by DNA from Wicky Thy^+, Sm-r 41 and the yield of transformants able to grow in M medium without thymidine was compared to the yield of streptomycin resistant transformants. Despite a strong tendency to revert toward prototrophy(especially W thy A2),the level of transformation was sufficiently high to show that both markers are integrated as efficiently as the marker Sm-r 41.

TABLE I

INTEGRATION EFFICIENCIES OF FUSIDIC ACID RESISTANCE MARKERS OF SPONTANEOUS ORIGIN

Clone number	WICKY			CHALLIS		
	Number of transformants Rif-rl $\times 10^4$/ml	Number of transformants Fus-r $\times 10^4$/ml	Fus-r / Rif-rl	Number of transformants Rif-rl $\times 10^4$/ml	Number of transformants Fus-r $\times 10^4$ /ml	Fus-r / Rif-rl
1	135	109	0.8	54	43	0.8
2	226	249	1.1	114	120	1.1
3	306	349	1.1	102	98	1.0
4	70	74	1.0	24	25	1.1
5	230	211	0.9	68	58	0.8
6	145	163	1.1	58	63	1.1
7	115	100	0.9	58	56	1.1
8	460	423	0.9	184	151	0.8
9	95	103	1.1	38	44	1.2
10	55	64	1.2	19	14	0.7

The sampling error in determining the number of transformants was less than 10%.

TABLE II

INTEGRATION EFFICIENCIES OF RIFAMPIN RESISTANCE MARKERS OF SPONTANEOUS ORIGIN

Clone number	Number of transformants Sm-r x 10^4/ml	Number of transformants Rif-r x 10^4/ml	Rif-r / Sm-r 41	Clone number	Number of transformants Sm-r x 10^4/ml	Number of transformants Rif-r x10^4 /ml	Rif-r / Sm-r 41
1	20	20	1.0	16	415	465	1.1
2	242	297	1.2	17	325	360	1.1
3	75	70	0.9	18	520	570	1.1
4	480	580	1.2	19	325	289	0.9
5	257	307	1.2	20	285	299	1.0
6	295	377	1.3	21	427	508	1.2
7	345	422	1.2	22	387	367	0.9
8	342	342	1.0	23	390	382	1.0
9	222	302	1.4	24	250	303	1.2
10	495	575	1.2	25	419	482	1.2
11	270	225	0.8	26	65	82	1.2
12	300	355	1.2	27	407	446	1.1
13	445	512	1.2	28	365	332	0.9
14	48	55	1.2	29	385	439	1.1
15	450	545	1.2	30	322	374	1.2

Since, in pneumococcal transformation, mutations induced by nitrous acid treatment of transforming DNA all fall into the low efficiency class when tested on discriminating strains (Sicard and Ephrussi-Taylor, 1965; Tiraby and Fox, in preparation), nitrous acid induced mutants were examined in streptococcal transformation. Mutants resistant to fusidic acid or kanamycin were isolated by the method described by Freese (1962), taking advantage of the linkage of these markers to Sm-r 41. Mutants were isolated by transformation of Wicky (rif-r) with nitrous acid treated Sm-r 41 DNA and challenging with streptomycin and either kanamycin or fusidic acid after having allowed 3 hours of growth for phenotypic expression. Doubly resistant colonies resulted from introduction by transformation of the Sm-r 41 marker and a mutation to resistance to either kanamycin or fusidic acid at a linked locus. Five such markers conferring kanamycin resistance and five conferring fusidic acid resistance were tested for their integration efficiencies, using rifampin resistance as a standard. These markers all manifested the same integration efficiencies as those that have already been described in Tables I and II.

(b) Heterospecific transformation

Streptococcus Challis. Chen and Ravin (1966) have decribed the behavior of a number of markers exhibiting variable efficiencies of integration in heterospecific transformation, e.g. using pneumococcal DNA to transform streptococcus. We have examined the behavior in heterospecific transformation of a number of pneumococcal markers whose characteristic integration efficiencies in the discriminating strains of pneumococcus are uniformly high in the non-discriminating mutants.

DNA's isolated from a number of Rifampin resistant mutants whose markers fall into various efficiency classes when tested on the discriminating pneumococcus strain R6, were tested on the Wicky strain of streptococcus. The relative yields of the two transformant classes are given in Table III. Here again the

TABLE III

INTEGRATION EFFICIENCIES OF PNEUMOCOCCAL RIFAMPIN RESISTANCE MARKERS IN STREPTOCOCCUS STRAIN WICKY

DNA number	R6 Rif-r / Sm-r 41	R6 Number of transformants Sm-r 41 x 10^3/ml	WICKY Number of transformants Rif-r x 10^3/ml	WICKY Rif-r / Sm-r 41
1	0.12	660	315	0.48
2	0.16	685	385	0.56
3	0.20	560	425	0.64
4	0.22	175	150	0.87
5	0.25	580	255	0.44
6	0.56	620	155	0.25
7	0.60	550	320	0.58
8	0.68	716	505	0.71
9	0.94	215	60	0.28
10	0.98	445	347	0.78
11	1.00	488	395	0.81
12	1.12	500	270	0.54
13	1.48	95	36	0.38
14	1.50	614	565	0.92
15	1.65	365	170	0.47

ratios have an uncertainty of about ±15%. There appears to be some variability over a range of about 3-fold but no correspondence between the efficiencies manifested in the discriminating strain in pneumococcus and the variability exhibited in the Wicky strain of streptococcus.

In an additional set of experiments a number of fusidic acid resistance markers were examined (Table IV). These markers include a number that resulted from the action of hydroxylamine and nitrous acid on DNA as well as several of spontaneous origin. The markers arising from mutagenesis are of the lower efficiency class as is common for mutants derived in this manner. Those of spontaneous origin are of the higher efficiency classes. Here again there is perhaps as much as a two- or threefold range of variation in the frequency with which markers may be donated in transformation but this variation fails to bear any obvious relationship to the efficiency classes within which the markers reside. The efficiencies of any individual marker seem similar in the two strains of streptococcus.

It should also be noted that markers derived by rather specific mutagenesis, in particular hydroxylamine, should always involve the same nucleotide alteration. The variation in the efficiencies of these markers suggests that the neighboring nucleotide sequence may play a role in dictating the frequency with which markers may be donated in heterospecific transformation. In addition it is apparent that the relative frequency with which markers conferring fusidic acid resistance are donated is substantially higher than those observed for the rifampin resistance markers. Such differences are often interpreted as resulting from different degrees of homology or persistence of polynucleotide sequence between heterologous strains(Chen and Ravin, 1966).

The markers conferring fusidic acid resistance are linked to loci conferring streptomycin resistance and might constitute mutations in a ribosomal protein. The efficiency with which fusidic acid resistance can be transferred in heterospecific transformation could be the result of the persistence in evolution of the genetic sequences responsible for ribosome synthesis (Dubnau et al., 1965).

4. DISCUSSION

An examination of more than 80 independent streptococcus mutants reveals no evidence of marker discrimination. There have been reports of a few streptococcal markers exhibiting low characteristic integration efficiencies (Chen and Ravin, 1966 ; Leonard, Coorley and Cole, 1966). It seems likely that these reduced efficiencies are the result of multi-site mutations, similar to those observed in pneumococcus, requiring extensive displacement of a polynucleotide sequence for marker integration.

Although there is evidence of variability in the integration efficiencies of various pneumococcal markers in heterospecific transformation of streptococcus, this variability does not reflect the features characteristic of the discriminating strains of pneumococcus. It would seem reasonable to assume that integration efficiencies reflect local similarities or differences in nucleotide sequences between pneumococcus and Group H streptococcus.

ACKNOWLEDGEMENTS

This work was supported by grant number AI 05388-10 from the U. S. Public Health Service and G.T. was partially supported by a fellowship from NATO.

TABLE IV

INTEGRATION EFFICIENCIES OF PNEUMOCOCCAL FUSIDIC ACID RESISTANCE MARKERS

Origin of mutants	R6	WICKY			CHALLIS		
	Fus-r / Sm-r 41	Transformants Sm-r 41×10^3/ml	Transformants Fus-r $\times 10^3$/ml	Fus-r / Sm-r 41	Transformants Sm-r 41×10^3/ml	Transformants Fus-r $\times 10^3$/ml	Fus-r / Sm-r 41
NH_2OH-induced							
1	0.08	120	396	3.3	135	320	2.4
2	0.14	180	270	1.5	205	348	1.7
3	0.13	265	490	1.8	340	653	1.9
4	0.12	189	535	2.8	95	285	3.0
5	0.15	294	436	1.5	230	435	1.8

HNO_2 induced							
6	0.12	134	330	2.5	145	305	2.1
7	0.15	87	295	3.4	60	170	2.8
8	0.16	420	456	1.1	705	1200	1.7
9	0.20	159	323	2.0	525	925	1.8
10	0.20	296	352	1.2	405	625	1.5
Spontaneous							
11	1.30	129	352	2.7	265	585	2.2
12	0.68	303	575	1.9	365	630	1.7
13	1.23	298	524	1.8	305	550	1.8
14	0.92	190	346	1.8	55	100	1.8
15	0.93	293	352	1.2	250	485	1.9

5. REFERENCES

CHEN, K. C. & RAVIN, A. W. (1966). J. Molec. Biol. 22:123.

DUBNAU, D., SMITH, I., MORELL, P. & MARMUR, J. (1965). Proc. Nat. Acad. Sci. U.S. 54:491.

EPHRUSSI-TAYLOR, H., SICARD, A. M. & KAMEN, R. (1965). Genetics 51:455.

FOX, M. S. (1957). Biochim. Biophys. Acta 26, 83

FREESE, E. & STRACK, H. B. (1962). Proc. Nat. Acad. Sci. U.S. 48:1796.

FRIEDMAN, L. R. & RAVIN, A. W. (1972). J. Bacteriol. 109:459.

GRAY, T. C. & EPHRUSSI-TAYLOR, H. (1967). Genetics 57:125.

GUERRINI, F. & FOX, M. S. (1968). Proc. Nat. Acad. Sci. U.S. 59:1116.

GURNEY, T. & FOX, M. S. (1968). J. Molec. Biol. 32:83.

KUWANO, M., SCHLESSINGER, D., RINALDI, G., FELICETTI, L. & TOCCHINI-VALENTINI, G. P. (1971). Biochem. Biophys. Res. Commun. 42:441.

LACKS, S. (1966). Genetics 53:207.

LACKS, S. (1970). J. Bacteriol. 101:373.

LEONARD, G. G., CORLEY, D. & COLE, R. M. (1966). Science 3726: 1255.

MORSE, H. G. & LERMAN, L. S. (1969). Genetics 61:41.

RANHAND, J. M. & COLE, R. M. (1971). J. Bacteriol. 106:712.

RANHAND, J. M., LEONARD, G. G. & COLE, R. M. (1971). J. Bacteriol. 106:257.

SICARD, A. M. & EPHRUSSI-TAYLOR, H. (1965). Genetics 52:1207.

SIROTNAK, F. M. & HACHTEL, S. L. (1969). Genetics 61:293.

WEHRLI, W. & MATTHYS, S. (1971). Bacteriol. Reviews 35:290.

INTEGRATION EFFICIENCIES IN DNA INDUCED TRANSFORMATION OF PNEUMOCOCCUS.GENETIC STUDIES OF MUTANTS INTEGRATING MARKERS WITH THE SAME EFFICIENCY

J. G. TIRABY*

J. P. CLAVERYS

A. M. SICARD

Laboratoire de Génétique, Université Paul Sabatier
Toulouse, France

ABSTRACT

Genetic studies in Diplococcus pneumoniae have shown that markers do not transform with the same efficiency. At least two classes have been defined : high efficiency markers (HE) and low efficiency markers (LE).The recombination process is somehow different for these two classes. We have shown that L.E. markers require, for an integration excision and repair mechanism whereas H.E. markers do not.

Using a method of transformation on solid medium especially adapted for pneumococcus we isolate not only transformation deficient strain, but also mutants modified in their recombination properties. A strain showing identical integration efficiency for HE or LE markers is more especially investigated; a genetic analysis has been conducted to be sure that only one gene controls this phenotype. The determination of the mode of transmission of acquired markers and the measure of the DNA uptake during transformation indicate that this strain integrates all markers according to the high efficiency pro-

* Present adress: Department of Biology, Massachusetts Institute of Technology, Cambridge, Mass. 02139. U.S.A.

cess i.e. without the excision repair mechanism. No other phenotypic properties could be found for this strain. Another strain showing similar efficiency for all markers due to a mutation to a second gene is described. This strain is radiation sensitive.

1. INTRODUCTION

Bacterial transformation requires several steps, starting with the binding of free DNA by the competent cells and ending with the alteration of the recipient genome. In terms of biochemical events, only some of the steps determining the fate of the DNA after uptake are characterized in *Diplococcus pneumoniae*. The nature of the physiological changes which occur during the development of competence is not completely understood. Also poorly understood in pneumococcal transformation is the nature of the process capable of discriminating between various markers. The most striking feature emerging from all the genetic studies in *Diplococcus pneumoniae* is the difference in the efficiencies with which markers representing point mutations are transferred by transformation. According to their individual integration efficiencies, auxotrophic markers or those conferring resistance to antibiotics can be divided into several classes. Two classes, one corresponding to high efficiency and the other to low efficiency, were found for mutations in the ami-A locus (Ephrussi-Taylor, Sicard and Kamen, 1965; Gray and Ephrussi-Taylor, 1967) and for mutants requiring uracil (Morse and Lerman, 1969), whereas four classes were found for mutations in the amylomaltose locus (Lacks, 1966) and for mutations in the structural gene of dihydrofolate reductase (Sirotnak and Hachtel, 1969). Regardless of the exact number of efficiency classes, two classes have been especially considered, the first with high efficiency (HE), the second with low efficiency (LE). These two classes differ in efficiency by about a factor of 10. Results from Lacks (1965); Ephrussi-Taylor (1966, 1968); Louarn and Sicard (1968, 1969) suggest that the integration mechanism of LE markers is different from that of HE markers. Let us summarize what we have already shown:

1) The genetic information is transferred to both recipient DNA strands for low efficiency markers whereas only one strand is modified at the time of integration of high efficiency markers. This conclusion is the result of four independent genetic observations: clonal analysis of individual colonies undergoing transformation (Louarn and Sicard, 1968); frequency analysis of double transformants for unlinked markers (Louarn and Sicard, 1969); measurement of the time for phenotypic expression of high and low efficiency markers; and kinetic analysis of the duplication for these markers after transformation. (Louarn, 1970).

2) The transmission to both strands of recipient DNA for LE markers has been interpreted as the result of an excision repair process so that the single stranded exogenous piece of DNA is integrated and copied. The recipient chromosome then becomes homoduplex (Ephrussi-Taylor, 1966).

3) The involvement of a DNA synthesis for integration of LE markers is suggested by experiments involving inhibition of this synthesis. If during transformation we block DNA synthesis, either by aminopterin which inhibits dihydrofolate reductase and therefore production of thymidylic acid or by thymidine starvation of competent thymidine - requiring pneumococcal cultures, the efficiencies of the LE markers decrease by a factor of two to three.

4) The time required for the inferred excision and repair process is very short since, to be effective, DNA synthesis must be blocked before the transforming DNA is added. If aminopterin is added to the competent culture at the same time or later than the transforming DNA, the efficiency of LE markers is unchanged (Sicard and Louarn, in preparation).

5) Low efficiency markers are single site mutations. The decrease of their transformability results from the confrontation of this mutated base in opposition to the wild type sequence (Sicard and Ephrussi-Taylor, 1966). These mismatched bases are somehow a block for recombination enzymes. The excision process could be limited to these unpaired bases. However this is not the case. Genetic experiments involving revertant strains as recipients indicated that the lenght of the excised fragment is obout the third to one half the size of a gene i.e. a few hundred pairs of bases. (Louarn, 1970).

In order to study the nature of the transient event occurring during pneumococcal transformation and the nature of the mechanism of discrimination between markers, attempts were made to isolate transformation-deficient mutants as well as mutants whose behaviour in relation to the integration efficiency of markers is altered. With this in mind, a method of the selection, based on the fact that pneumococcus can be transformed on solid medium, was established. This method which will be described in another paper proved to be efficient for isolating both transformation-deficient mutants and recombination-modified mutants.

2. ISOLATION OF MUTANTS

(a) Mutants modified in their transforming ability

Usually 1-2 per cent of the colonies obtained from a culture treated with nitrosoguanidine failed to give optochin and streptomycin transformants. Approximately 30 mutants which did not transform or which transformed poorly for both HE and LE markers were isolated from five different treated cultures. In a preliminary attempt to classify these mutants we determined their sensitivity to ultraviolet light. According to their response to U.V., three classes can be distinguished; some strain were respectively 2.5 and 3 times more resistant than the standard wild type; one strain had the same sensitivity as the wild-type; two others were more sensitive than C13.

In order to determine the nature of the deficient steps, the transformation-deficient mutants, were allowed to react with DNA from a streptomycin resistant strain labelled with tritiated thymidine. The DNA failed to enter for some strains. Strain 405 yielded transformed cells, but fewer than the wild-type. The decrease in the number of transformed bacteria in this strain is partially due to a decrease in the integration of the transforming DNA into the genome of the cells, since the number of transformants per count of radioactivity incorporated was five times smaller for 405 than it was for C13.

(b) Mutants modified in their recombination ability

Mutants of great interest for studying recombination in pneumococcus are those which would modify the integration efficiency of HE and/or LE markers compared to the wild type strain. A few such recombination-modified mutants were isolated by the method of transformation on plates.

Mutants T6 and T16 represent a class of recombination-modified mutants. In these strains the efficiencies of the LE markers only are lowered 3-4 fold compared to the wild type strain (Table I). In addition, the delay in the complete phenotypic expression of the LE markers after transformation is greatly increased.

In strain 500 the variation of the integration efficiency for various markers is less pronounced than in C13 (Table I). From the measure of the DNA uptake it appears that only the efficiencies of HE markers are modified in this strain (Table II). The efficiencies of LE markers are the same as in the wild type, but the efficiency of the streptomycin-resistant marker is lowered. The data in Table I imply that another HE marker, ami A-r1, behaves like str-r 41, since the efficiencies of both markers are the same when tested in strain 500.

A class of mutants in which LE marker efficiencies increase to almost the level of HE marker efficiencies is already known. Such mutants were isolated fortuitously many years ago : strain Rx (Ravin, 1959), strain 84 (Litman, 1961) and more recently the Hex strains (Lacks, 1970). We have also isolated such a mutant, strain 401, which shows no difference in efficiencies between markers (Louarn, Tiraby and Sicard, 1970). The properties of this mutant will be described in the next section. By measuring the uptake of labelled DNA it will be shown that, in contrast to strain 500, 401 increased the efficiency of only the LE markers. In addition, 401 and also Rx (Fox, personal communication) and Hex-1 (Lacks, 1970) have the same level of resistance to ultraviolet light as the wild type strain, whereas the mutation which modifies the efficiencies of markers in strain 500 confers sensitivity to radiation.

TABLE I

INTEGRATION EFFICIENCY OF MARKERS AMI A-R1, AMI A-R9 AND OPT-R2 IN RECOMBINATION-MODIFIED MUTANTS AND IN THE WILD-TYPE STRAIN.

Markers	Strains			
	Cl_3	500	T_6	401
ami A-r1	1.2 ± 0.1	1.09 ± 0.1	1.2	0.82 ± 0.03
ami A-r9	0.15 ± 0.05	0.46 ± 0.05	0.05	0.82 ± 0.03
opt-r2	0.18 ± 0.05	0.39 ± 0.06	0.05	0.48 ± 0.04

The integration of a given marker is determined by the ratio of the number of transformants for this marker over the number of transformants for the reference marker conferring resistance to streptomycin (str-r 41). The different strains were transformed by DNA from strain 69 and by DNA bearing both markers str-r41 and ami A-r 1. The competent cells were allowed to react with DNA for 20 minutes, then the cells were plated in a non-selective medium. After 150 minutes incubation at 37 C for strains C13, 500, 401, and 300 minutes incubation for strain T6, the transformants were scored by pouring a second layer of medium containing the antibiotic over the first layer.

TABLE II

DNA UPTAKE AND TRANSFORMATION EFFICIENCY IN STRAINS C13 (WILD TYPE) AND 500 (RECOMBINATION-MODIFIED)

Strains	DNA concentration μg/ml	(1) Number of cocci in colony-forming unit	(2) Counts/min/ml of culture	(3) Number of transformants per ml of culture		(4) Normalized transformation efficiency	
				Streptomycin x 10^4	Optochin x 10^3	Streptomycin	Optochin
Wild type	0.25	2.1	14930	313	563	1.00	1.00
	0.25	2.1	10370	243	437		
(C13)	0.05	2.1	2510	57	103		
500	0.25	2.7	17500	87.5	341		
	0.25	2.7	9850	74.5	291	0.35	0.92
	0.05	2.7	2820	23	90		

The measurements were made as described in Materials and Methods. The average number of cocci in the chain-growing competent pneumococcus were determined by direct count in a Petroff-Hauser chamber. The normalized transformation efficiency is defined as

$$\frac{(2)\ \text{Wild type} \times (3)}{(2) \times (3)\ \text{Wild type}} \times \frac{(1)\ \text{Wild type}}{(1)}$$

The values of the normalized transformation efficiency are the averages of the values obtained with the three different experiments reported in this table.

The method of transformation on Petri dishes provides a means of isolating both transformation-deficient mutants and recombination-modified mutants. Recombination-modified mutants are those in which the pattern of integration of only one class of markers is altered compared to the wild type. They are of special interest and the next section will be devoted to the study of one of them.

3. THE STRAIN 401

(a) Isolation

Using nitrosoguanidine as the mutagenic agent we have isolated one clone which transforms one He marker (str-r 41) and one LE marker (opt-r 2) with the same efficiency. Since it is likely that, under our experimental conditions, several mutations might have been induced per cell, the new character has been transferred into our standard wild type strain, clone 3, by transformation. Besides the integration efficiency, no other difference can be detected between the mutant and the wild type, including the generation time at 25° C or 37° C. Therefore this characteristic cannot be easily screened. We have therefore taken advantage of the ability of pneumococcus to give rise to transformed bacteria on solid medium. Wild type cells were transformed by DNA extracted from the mutant strain and clones showing the same efficiency of transformation were selected by the method of transformation on solid medium. One clone, named 401, isogenic to clone 3 except for the selected mutation, was isolated. The transmission of this mutation from the original strain to the wild type suggests that only one gene is involved in the 401 character.

(b) Integration efficiency of various markers

The property of strain 401 to transform one LE and one HE marker with the same efficiency was extended to various markers. Two markers belonging to the same locus, one HE (ami A-r 1) and the other LE (ami A-r 9), displayed the same integration efficiency. The integration efficiency of the opt-r 2 marker was never found to be as high as the reference str-r 41 marker. Other LE markers of spontaneous origin, conferring resistance to 5 μg of optochin were isolated. All the 9 different markers tested gave the characteristic integration efficiency of the opt r-2 mutation. Likewise the integration efficiency of 15 LE ami A markers was found to be 0.80 with no more than 10% variation from this value (Table III).

(c) Radiation sensitivity

Strain 401 has the same sensitivity to UV light as the wild type (Fig. 1). It seems, therefore, that the modification introduced by the mutation 401 on the efficiency of transformation has no effect on sensitivity to radiation. In order to uncover such interference between these properties, a gene conferring sensitivity to UV was transferred into C13 (strain 402) and 401 (strain 406). Doses sufficient to reduce survival to 37% were 155 ergs/mm^2 for C13 and 401 and 10 ergs/mm^2 for 402 and 406. Thus sensitivity to UV light is independent of the effect on transformation efficiency controlled by gene 401.

TABLE III

INTEGRATION EFFICIENCY OF VARIOUS MARKERS IN STRAIN 401 AND C13

Recipient strains	markers				
	ami A-r1 (HE)	ami A-r9 (LE)	opt-r2 (LE)	ery-r6 (HE)	ery-r2 (HE)
C13	1.10 1.22	0.15 0.13	0.20 0.16	0.60 0.60	0.65 0.60
401	0.83 0.85 0.80	0.80 0.82 0.85	0.45 0.48 0.50	0.90 0.95 0.95	0.90 1.00 0.90

The integration efficiency of a given marker is determined by the ratio of the number of transformants for this marker over the number of transformants for the reference marker conferring resistance to streptomycin (str-41). Strains C13 and 401 were transformed by DNA from strain 69 and by DNA bearing both markers <u>str-41</u> and <u>ery-r2</u> .

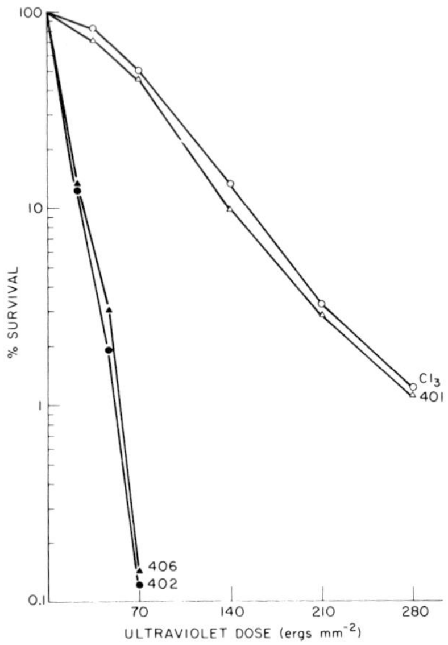

FIG. 1 - THE FRACTION OF COLONY-FORMING UNITS SURVIVING AS A FUNCTION OF THE ULTRAVIOLET RADIATION DOSE

The wild-type strain C13 (o), a mutant that gives only high integration efficiency, 401 (△), an ultraviolet sensitive mutant, 402 (●), and a mutant ultraviolet sensitive that gives only high integration efficiency, 406 (▲), were tested for survival after ultraviolet irradiation.

X-ray sensitivity of 401 was compared to the sensitivity of the wild type strain and was found identical for both strains (Fig. 2).

Genetic markers of transforming DNA are inactivated by U.V. irradiation. Ephrussi-Taylor, Sicard and Kamen (1965) observed that the sensitivity to U.V. is higher for LE markers than for HE markers belonging to the ami A locus. This inverse

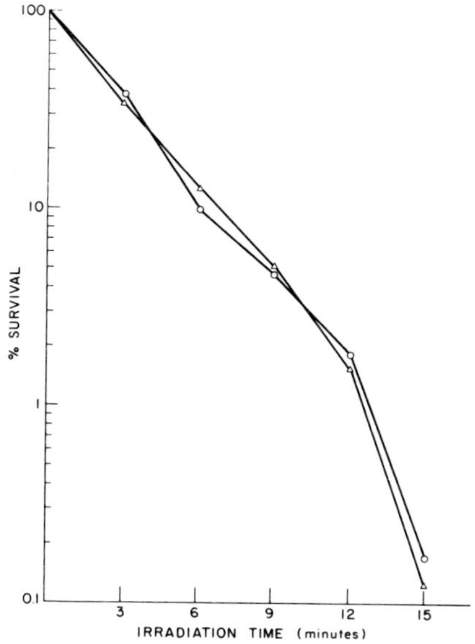

FIG. 2 - THE FRACTION OF COLONY-FORMING UNITS SURVIVING AS A FUNCTION OF THE X-RAYS DOSE

correlation between U.V. sensitivity and integration efficiency was confirmed for the different efficiencies of mal markers (Lacks, 1965), and methicillin resistance markers (Hotchkiss and Gabor, 1970). We thought it of interest to compare the sensitivity of various markers using 401 and wild type strain as recipients. When a solution of DNA is U.V. irradiated the loss of transforming activity of a given marker depends also on the molecular weight of the DNA (Marmur et al.,1961). In order to avoid the usual variation in the molecular weight of different preparations of DNA, one single solution of DNA was used. The DNA of strain 69 bearing several markers was irradiated at different doses and used to transform both strains C13 and 401. The results of one experiment are given in Fig. 3.

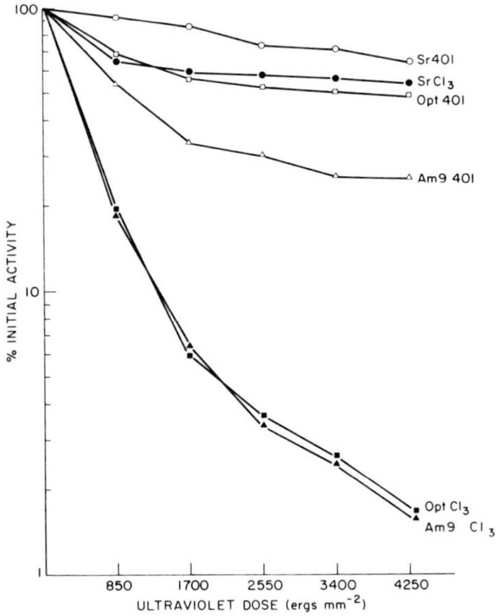

FIG. 3 - INACTIVATION OF MARKERS STR-R 41, OPT-R 2, AND AMI A-R 9 AFTER ULTRAVIOLET IRRADIATION OF THE DNA IN SOLUTION

The recipient strains were C13 (filled symbols) and 401 (open symbols).

The low efficiency markers such as opt-r2 or ami A-r9 exhibit a lower U.V. sensitivity in strain 401 than in C13. Thus in strain 401, LE markers increase their efficiency of transformation and at the same time lose their U.V. sensitivity. Moreover the reference marker str-r 41 is more resistant in strain 401 than in C13. If we assume that the inverse correlation between U.V. sensitivity and integration efficiency is correct, the basic efficiency of str-r 41 marker will be higher in strain 401 than in C13. Indeed this was verified by the following experiment.

(d) Measure of DNA uptake during transformation

Strains 401 and C13 were transformed by DNA labelled with tritiated thymidine. The amount of transforming DNA incorporated into the genome of the recipient bacteria was determined by the ratio of the number of transformants str-r 41/ml to the number of counts/ml.

The basic transformation efficiency for the reference marker str-r 41 increased by a factor of 1.5 to 1.6 in 401 as compared to the wild type strain (Table IV). This increase corresponds to the difference existing between the efficiencies of HE markers and markers of very high efficiencies as found by Lacks (1965) and Sirotnak and Hachtel (1969). Mutants in the ami A locus with very high efficiency were also recently isolated in our laboratory (Tiraby, Sicard to be published). Thus the information carried by HE markers is probably partly lost when C13 is used as recipient. Since we have shown that LE markers are integrated in strain 401 almost as well as HE markers, the basic integration efficiency is increased for both types of markers in this strain.

(e) Mode of integration of the markers

Measurement of the uptake of radioactive DNA has shown that the integration efficiency of str-r 41 in strain 401 is as

TABLE IV

DNA UPTAKE AND TRANSFORMATION EFFICIENCY IN STRAINS C13 (WILD-TYPE) AND 401 (RECOMBINATION-MODIFIED)

Recepient strains	DNA concentration µg/ml	(1) Counts/min/ml of culture	(2) Number of streptomycin transformants/ml of culture x 10^4	Normalized transformation efficiency
C13	0.25	14930	313	
	0.25	10370	243	1.00
	0.05	2510	57	
401	0.25	13120	450	
	0.25	8750	335	1.5
	0.05	2530	83.4	

The average number of cocci per colony-forming unit was the same for both recipient strains. The normalized transformation efficiency is defined as (2)/(1). The values of the normalized transformation efficiency are the averages of the values obtained with the three different experiments reported in this table, the efficiency of C13 being taken as reference.

high as that of the very high efficiency markers in the wild type strain.Gray and Ephrussi-Taylor (1967) have postulated that the efficiencies of both HE and LE markers could be explained by the involvement of a mechanism of excision and repair.

A precise genetic study of the pattern of transmission of genetic information in this strain was undertaken to specify the true mode of integration of markers in strain 401.

1) Determination of the frequency of double transformants for strains 401 and C13.

Louarn and Sicard (1969) showed that the frequency of doubly transformed bacteria is characteristic of the mode of integration of the markers. When C13 is transformed,the genetic information of HE markers is transmitted into either recipient strand at random, whereas LE markers modify both recipient strands. The frequency of double transformants for a pair of LE-LE markers will be higher than that for a pair of HE-HE markers.Thus the value R of the ratio R = D xN/A x B will depend on the efficiencies of the selected markers (where N is the number of colony forming units per ml; D, the number of doubly transformed bacteria for the independent markers A and B;A and B, the number of singly transformed bacteria for these markers). These values were determined for several markers in each of the two strains, C13 and 401.

The DNA- bacterial complexes giving an efficiency of transformation of at least 10% were used for these experiments. This high frequency of transformation can be reached only if all the bacteria in the culture are competent. The double transformants were isolated using the same concentration of antibiotics as for selection of single transformants. The R values in three different experiments with C13 and four experiments with 401 are given in Table V. In strain 401, the values obtained for the three pairs of markers under consideration were not only identical with each other but were also identical with the value found for the pair of HE-HE markers,ami A rl - str-r41 in strain C13. Therefore all markers are transformed in strain 401 by the same recombination mechanism as high efficiency markers in strain C13.

TABLE V

R VALUES FOR VARIOUS PAIRS OF MARKERS IN THE TRANSFORMATION OF STRAINS C13 AND 401

Recipient strains	Pairs of markers		
	ami A-r1, str-r41 (HE-HE)	opt-r2, str-r41 (LE-HE)	opt-r2, ami A-r1 (LE-HE)
C13	0.28	0.38	0.39
	0.34	0.45	0.45
	0.37	0.46	0.48
	0.33	0.43	0.44
401	0.29	0.30	0.29
	0.35	0.35	0.33
	0.38	0.39	0.36
	0.38	0.40	0.38
	0.35	0.36	0.34

The reported values of R corresponding to the ratio DN/AB were computed from three independent experiments for C13 and four for 401. DNA from strain 39 bearing str-r41. opt-r2,ami A-r1 markers was used.

2) Kinetics of appearence of transformed bacteria

The study of the kinetics of the appearance of transformed bacteria can also be used to identify the process of integration: both recipient strands of C13 are modified during transformation by LE markers. The information is transcribed into messenger RNA from one of the two DNA strands.However, only one recipient strand chosen at random is modified by HE markers. The full expression of this information requires an additional replication of the DNA (Ephrussi-Taylor, 1966). Therefore the maximum number of transformed bacteria is obtained sooner for LE markers than for HE markers.The kinetics of appearance of the phenotypically expressed opt-r2 and str-r41 markers was determined in both strains C13 and 401.

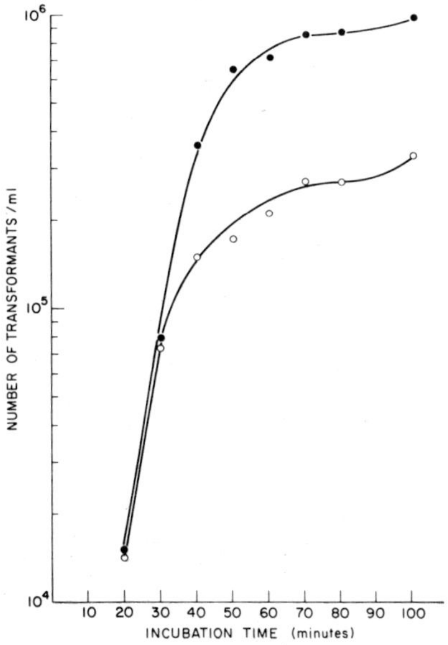

FIG. 4 - PHENOTYPIC EXPRESSION OF A HIGH EFFICIENCY MARKER IN WILD-TYPE AND MUTANT STRAINS

The kinetics of appearance of transformants str-r41 were determined in C13 (open symbols) and in 401 (filled symbols).

The time at which a number corresponding to 50% of the str-r41 transformants was reached was approximatively the same for both strains (Fig. 4), whereas the opt-r2 transformants were expressed 15 to 20 min later in strain 401 than in strain C13 (Fig. 5). This result suggests that low efficiency markers are integrated in strain 401 in the same way as HE markers are integrated in strain C13.

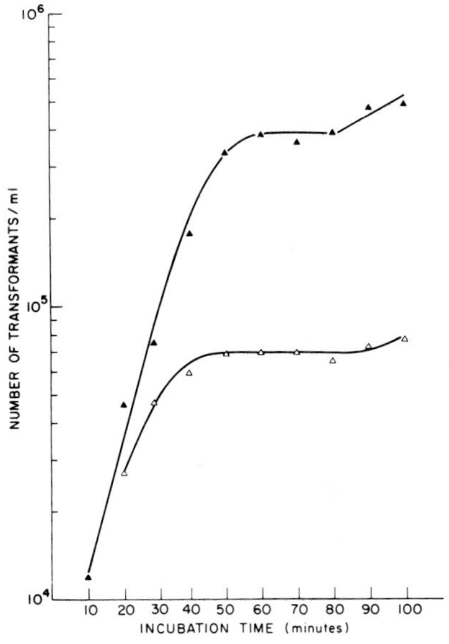

FIG. 5 - PHENOTYPIC EXPRESSION OF A LOW EFFICIENCY MARKER IN WILD-TYPE AND MUTANT STRAINS

The kinetics of appearance of transformants <u>opt-r2</u> were determined in C13 (open symbols) and in 401 (filled symbols).

4. KINETICS OF DIVISION OF THE TRANSFORMED BACTERIA

As a consequence of the different modes of transmission of the genetic information, the number of transformed bacteria for a LE marker will increase after incubation sooner than the number of transformed bacteria for an HE marker in the wild-type strain (Louarn 1970).

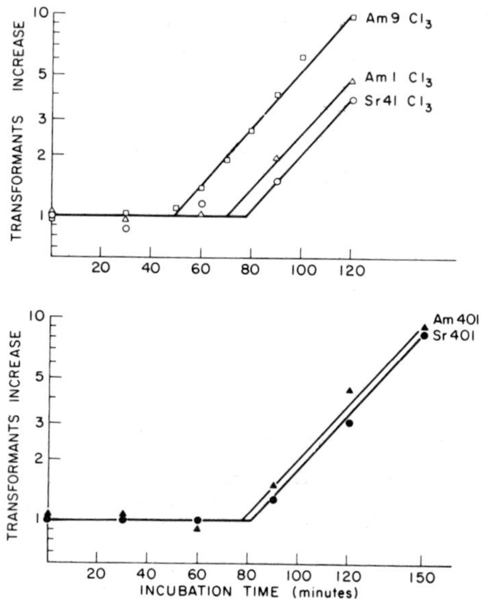

FIG. 6 - MULTIPLICATION OF TRANSFORMANTS FOR VARIOUS MARKERS IN WILD-TYPE AND MUTANT STRAINS

The wild-type strain (C13) was transformed by DNA from strains 69 and 39. The mutant strain (401) was transformed by DNA from strain 39. The onset of division for markers ami A-r9 (LE), ami A-r1 (HE), and str-r41 (HE) in C13 and for markers amiA-r1 (HE), and str-r41 (HE) in 401 was determined as described.

The kinetics of division of the transformed cells for

both markers was determined in C13 and 401. The experimental method used was very similar to the one used in the experiments above, except that the transformed bacteria were allowed to express their phenotype before being scored. It can be seen in Fig. 6 that the LE mutation ami A-r9 began to multiply in the recipient wild type strain before the HE mutation ami A-r1 belonging to the same locus. The time separating the two onsets of division for the two kinds of markers was about 24 minutes, which corresponds to one generation in the yeast extract - neopeptone medium. The time of onset of division in the mutant strain was the same for the two HE markers tested and corresponds to that found for the HE marker in strain C13.

Thus, the experimental results are on the whole in favour of a single mode of integration of the markers in strain 401 which is identical to that of the HE markers in strain C13.

5. CONCLUSION

As a result of a mutation in one gene, strain 401 has lost the ability to eliminate most of the information carried by low efficiency markers during transformation. The mechanism which in the wild type is able to discriminate between the two classes of markers has disappeared. The sensitivity of this strain to radiation is normal. However, for all markers tested, U.V.-irradiated DNA is more resistant. Measurement of the amount of incorporated radio-active transforming DNA indicates that transformation efficiency in strain 401 is higher than in the wild type. It corresponds to the highest efficiency class found by Lacks (1966), and Sirotnak and Hachtel (1969). Thus, it appears that a fraction of the HE markers whose efficiency is

1 compared to 1.6 for the very high efficiency mutations, must be excluded from the wild type.

The mode of integration in the mutant 401 was shown to be identical for all the various markers tested, contrary to the wild type strain, where the occurrence of two modes was suggested (Ephrussi-Taylor, 1966; Louarn and Sicard, 1968, 1969). The mode of integration is one that acts for the HE markers. Thus, our results suggest that the correcting mechanism implied in the integration of LE markers in strain C13 does not function in strain 401. This is in agreement with Peterson and Guild's conclusions (1968) about strain Rx which is similar to 401 with respect to marker efficiencies; they found that all markers are integrated into either strand of the recipient DNA of strain Rx.

The best model to account for the present results is that originally proposed by Ephrussi-Taylor (1966): a system of correction acting by an excision and repair synthesis for the integration of LE markers. During transformation, when pairing between the transforming and recipient strand is not permissive due to mismatched bases, a process of correction excises one of the two strands at the level of the mutation, then resynthesizes this strand using the complementary strand as a template. The exogenous strand is preferentially eliminated about ten times more frequently than the endogenous strand. The heteroduplex formed with an HE marker is a permissive confrontation and is not, or only slightly, subject to excision.

Applying the model proposed by Ephrussi-Taylor, the excision enzyme specific for mismatched bases should be inactive in the strain 401. If this model is correct, such an enzyme should be different from those involved in the process of DNA repair by excision after U.V. irradiation, since strain 401 has the same sensitivity to U.V. as strain C13. This is in agreement with our previous work, showing that inhibition of repair enzymes by caffeine did not modify the characteristic efficiency of markers (Louarn and Sicard, 1965). Experiments are now in progress to test these hypotheses and to look for the inferred nucleases.

The other mutation (strain 500), lowering the integration efficiency of the LE markers and conferring sensitivity to radiations, is also of special interest and is at present, under active investigation.

6. REFERENCES

EPHRUSSI-TAYLOR, H. (1966). Genetics 54:211.

EPHRUSSI-TAYLOR, H. (1968). Symposium of the VIIth Int. Cong. of Microbiology. Stockholm.

EPHRUSSI-TAYLOR, H., SICARD, A.M. & KAMEN, R. (1965). Genetics 51:455.

GRAY, T.C. & EPHRUSSI-TAYLOR, H. (1967). Genetics 57:125.

HOTCHKISS, R.C. & GABOR, M. (1970). Ann. Rev. Genetics 4:193.

LACKS, S. (1965). In KOHOUTOVA, M. & J. HUBACEK (ed.), Academia Prague, p. 159.

LACKS, S. (1966). Genetics, 53:207.

LACKS, S. (1970). J. Bact. 101:373.

LITMAN, R.M. (1961). J. Chimie Phys. 58:997.

LOUARN, J.M. & SICARD, A.M. (1965). Compt. Rendus. Ac.Sc.261:852.

LOUARN, J.M. & SICARD, A.M. (1968). Biochem. Biophys. Res. Comm. 30:683.

LOUARN, J.M. & SICARD, A.M. (1969). Biochem. Biophys. Res. Comm. 36:101.

LOUARN, J.M., TIRABY, G. & SICARD, A.M. (1970). "Intégration des marqueurs à forte et faible efficacité chez le pneumocoque". Société Française de Génétique 36-48. Centre d'Hémotypologie du CNRS. CHU Purpan, 31 Toulouse (Editors).

LOUARN, J.M. (1970). "Etude de la recombination génétique chez *Diplococcus pneumoniae*". Thesis,Université de Toulouse, France.

MARMUR, J., ANDERSON, W.F., MATTHEWS, L., BERNS, K., GAJEWSKA,E., LANE, D. & DOTY, P. (1961). J. Cell. Comp. Phys.58:33.

MORSE, H.G. & LERMAN, L.S. (1969). Genetics 61:41.

PETERSON, J.M. & GUILD, W.R. (1968). J. Bact. 96:1991.

RAVIN, A. (1959). J. Bact. 77:296.

SICARD, A.M. & EPHRUSSI-TAYLOR, H.E. (1966). Compt. Rendus. Ac. Science, 262:2305.

SIROTNAK, F.M. & HACHTEL, S.L. (1969). Genetics 61:293.

LINKAGE STUDIES WITH TWO TRANSFORMABLE STRAINS OF PNEUMOCOCCUS

L. O. BUTLER

Department of Medical Microbiology,
St George's Hospital Medical School,
Hyde Park Corner, London,
S.W.1, England

ABSTRACT

A start has been made to produce a general map of the pneumococcus chromosome in terms of the order of replication of the markers. This has been achieved by application of the density-shift method, followed by an analysis of the resultant density-gradients for the determination of the transforming activity for the markers concerned, using strain clone 3 (Cl3) and strains derived from it as recipients. At the same time, a search has been made for linkage groups. So far, this has revealed only one pair of linked markers, viz, the genes responsible for resistance to erythromycin (*ery-r2*) and tetracycline (*tet-A*). No linkage was found between *str-r41* and *ery-r2*, nor between *str-r41* and *tet-A*, although (a) on the replication map *str-r41* occurs between *ery-r2* and *tet-A*, and (b) linkage between *str-r53* and *ery-r2* has been described by Ravin (1967). This anomaly does not appear to be due to the method of preparation of the transforming DNA, and provides evidence for the presence of more than one replicon in the pneumococcus chromosome, and probably bi-directional replication.

It was thought that the differences observed concerning the erythromycin-streptomycin linkage could be due to the use of different recipient strains. Using the recipient strain of Ravin (SIII-I), and its homologous DNA carrying the *str-r53* and *ery-r2* markers, the linkage has been confirmed. However, no linkage is obtained in the heterologous cross with strain C13 as recipient, whereas in the heterologous cross with SIII-I as recipient, a number of markers are linked. The implications of these results will be discussed.

1. INTRODUCTION

The work I will describe has been carried out in my laboratory with the help of Mr. Martin Smiley, Mr.Geoffrey Nicholas and Miss Marion Gray.

A start has been made on the making of a general map of the pneumococcus chromosome, by plotting the relative position of the markers in the order of replication by means of the density - shift method, first utilised for *B. subtilis* by O'Sullivan and Sueoka (1967) and Dubnau, Goldthwaite, Smith and Marmur (1967). 5-Bu and D, given as D_2O, were used as the density markers. Briefly, a method was devised to prepare synchronised cultures of the pneumococci, growing in a medium designed to give the density shift, and samples from such cultures were then removed after varying incubation times. These were lysed, centrifuged in CsCl, and the resulting density gradients analysed for the distribution of the transforming activity for the markers concerned, using strain Clone 3 (C13) and strains derived from C13 as the recipient organisms. At the same time, a search for linkage groups was made by making dilution curves, scoring for single and double transformants, together with counts of colony-forming units made on control tubes not containing DNA.The dilution curves were plotted and the frequencies of transformation for each transformant class were determined. Those for the double transformants were compared to the calculated ratio expected as the maxima for unlinked systems, obtained by multiplying together the frequencies of the two single events. Pneumococcal competent cultures contain a high proportion of competent cells (Fox and Hotchkiss, 1957; Nester and Stocker, 1963; Porter and Guild, 1969), and hence a comparison of the frequency of transformation does not give rise to spurious linkage as described with B. subtilis (Anagnostopoulos and Crawford, 1961) due to the comparatively low number of competent cells in competent cultures of B. subtilis (Nester and Stocker, 1963; Cahn and Fox, 1968).

2. STRAIN C13 AS RECIPIENT

Our previous work has placed eleven markers on a replication map, which is shown in Fig. 1 (Butler and Smiley, 1972). In the present case, we are particularly interested in the region located towards the end of the replication map where the order is -opt-r2 - pab-B - amiA-rl - ery-r2 - str-r41 - tet-A- . Also linkage has been looked for between the genes conferring resistance to streptomycin, erythromycin, aminopterin, optochin and tetracycline, and has been found only between erythromycin and tetracycline, with no evidence of linkage between erythromycin and streptomycin, nor tetracycline and streptomycin (Butler and Nicholas, 1972). Table I gives examples of the frequency of transformation for the pairs tet-A opt-r2 , tet-A str-r41, tet-A amiA-rl and tet-A ery-r2, and it can be seen that linkage occurred between tet-A and ery-r2, but not with any of the other markers. Fig. 2 gives an example of the

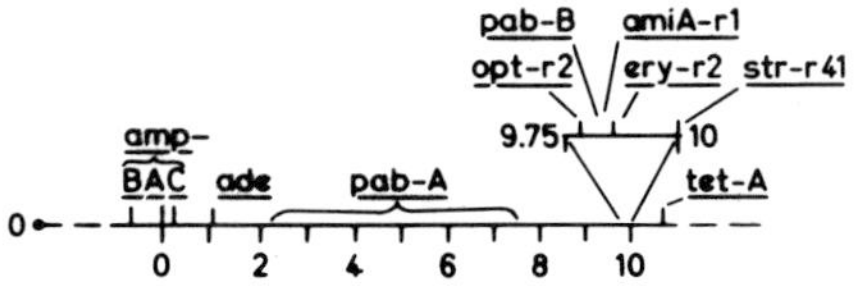

FIG. 1 - Map of the pneumococcal chromosome derived from density-shift experiments (from Butler and Smiley, 1972).

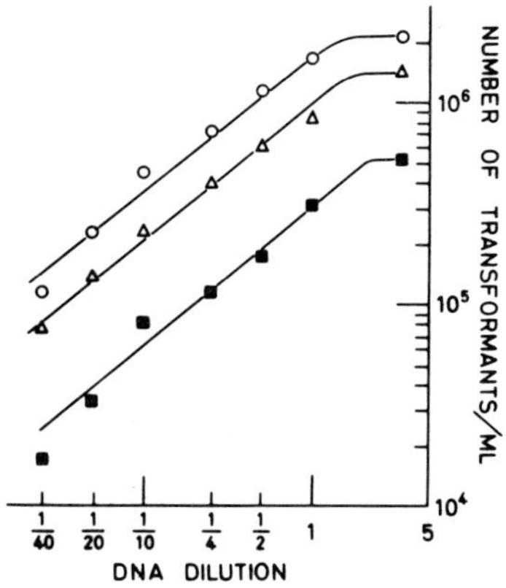

FIG. 2 - Dilution curves for the transformants to markers tet-A (O), ery-r2 (Δ), and tet-A ery-r2 (■), resulting from the cross Cl3 x ($Cl3\text{-}3Tc3r_2Q$) (from Butler and Nicholas, 1972).

plot of a dilution curve of the tet-A, ery-r2 and tet-A ery-r2 transformants.

These results are interesting for two reasons: (a) the replication map shows str-r41 occurring between ery-r2 and tet-A, and hence linkage would also have been expected with str-r41; and (b) Ravin (1966; Ravin and Chen, 1967) has described linkage between ery-r2 and str-r53. Since str-r53 is

TABLE I

DOUBLE TRANSFORMANTS INVOLVING THE <u>tet-A</u> MARKER. COMPARISON OF THE FREQUENCIES OF DOUBLE TRANSFORMANTS FOUND EXPERIMENTALLY TO THOSE CALCULATED AS MAXIMA FOR UNLINKED MARKERS

(From Butler and Nicholas, 1972)

Ratio $\frac{\text{Double transformants}}{\text{c.f.u.}}$ expressed as the reciprocal of the frequency											
tet-A opt-r2		tet-A str-r41		tet-A amiA-r1		tet-A ery-r2					
Calc.	Exptl.	Calc.	Exptl.	Calc.	Exptl.	Calc.	Exptl.	Calc.	Exptl.		
C13 x ($Tc3r_2SQ$)		C13 x ($Tc3Amlr_2SQ$)		C13 x ($Tc3Amlr_2SQ$)		C13 x ($Tc3Amlr_2SQ$)		C13 x ($Tc3r_2Q$)			
* 495	889	99	244	1,331	6,478	1,128	55	11,136	350		
640	1,143	156	388	1,404	16,556	2,040	127	24,753	590		
935	3,368	255	608	2,040	9,313	5,472	204	51,269	1,060		
1,680	5,818	422	1,406	2,310	24,833	11,324	362	138,699	1,630		
3,976	13,617	783	2,509	3,190	74,500	31,626	604	310,821	2,310		
7,718	17,778					77,070	1,646	1,086,168	5,670		
↓ 23,568	35,556					142,839	1,872	3,362,943	11,200		

* The arrow indicates decreasing concentrations of DNA.

situated in the same locus as str-r41, and may cover the latter marker, str-r41 would be expected to behave similarly.

TABLE II

DOUBLE TRANSFORMANTS INVOLVING THE tet-A MARKER, USING DONOR DNA PREPARED BY METHOD II. COMPARISON OF THE FREQUENCIES OF DOUBLE TRANSFORMANTS FOUND EXPERIMENTALLY TO THOSE CALCULATED AS MAXIMA FOR UNLINKED MARKERS. (From Butler and Nicholas, 1972).

Ratio $\frac{\text{Double transformants}}{\text{c.f.u.}}$ expressed as the reciprocal of the frequency

$C13(Tc3r_2S)_{II}$

	tet-A ery-r2		tet-A str-r41		ery-r2 str-r41	
	Calc.	Exptl.	Calc.	Exptl.	Calc.	Exptl.
* ↓	247	44	78	1067	114	421
	400	54	128	2286	200	889
	451	78	190	5704	290	1600
	1296	133	540	20000	960	4849
	2604	128	966	13913	1426	9697
	14240	386	5340	246150	9600	37210
	144120	970			57728	139130

* The arrow indicates decreasing concentrations of DNA.

It seemed possible that the method of preparation of the transforming DNA may have caused breakage of the chromosome at such a place to cause loss of the erythromycin-streptomycin linkage, but DNA prepared in a more gentle fashion still gave no linkage between ery-r2 and str-r41, but gave linkage between ery-r2 and tet-A (Butler and Nicholas, 1972), the results being

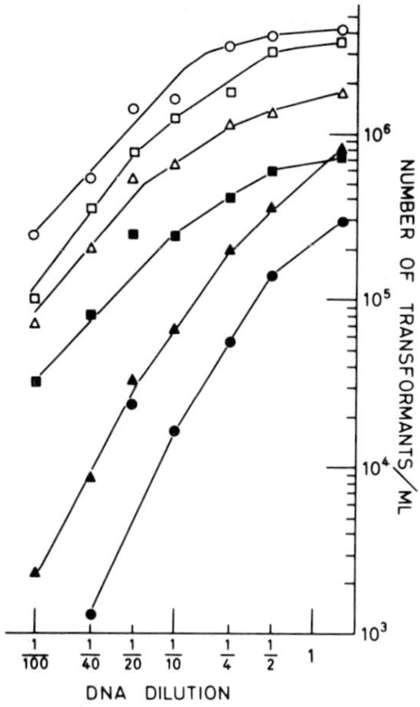

FIG. 3 - Dilution curves for transformants to the markers str-r41 (O), tet-A (□), ery-r2 (△), tet-A ery-r2 (■), str-r41 ery-r2 (▲, count x 10) and tet-A str-r41 (●, count x 10) resulting from the cross C13 x $(C13\text{-}3Tc3r_2S)_{II}$. (from Butler and Nicholas, 1972).

summarised in Table II and Fig. 3. These relationships were confirmed by clonal analysis (Butler and Nicholas, 1972). The anomalous linkage in relation to the replication therefore clearly shows that the replication map does not give the actual order of the genes on the C13 chromosome and gives evidence for the presence of more than one replicon, as illustrated in Fig. 4. Model (a) shows a circular chromosome divided into two replicons by having one initiation point from which replication proceeds in both directions. The markers ery-r2 and tet-A would be situated towards the end of one replicon, and str-r41 towards the end of the other replicon but at a point which would replicate after ery-r2 but before tet-A. Such bidirectional replication has been suggested by Caro and Berg (1968, 1969), Masters and Broda (1971), Yahara (1971) and Bird, Louarn and Caro(personal communication) for *Escherichia coli*, and by Nishioka and Eisenstark(1970) for *Salmonella typhimurium*.

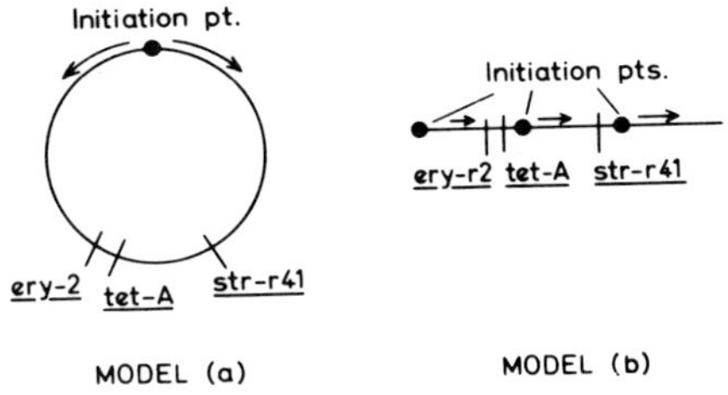

FIG. 4 - Models proposed for the replication of the pneumococcal chromosome. Model (a) having one initiation point but bidirectional replication giving two replicons. Model (b) having several initiation points with uni-directional (or bidirectional) replication in several replicons.(From Butler and Nicholas, 1972).

In Model (b), the chromosome, depicted for simplicity as a straight line, has been divided into several replicons each with its own initiation point. Replication could be uni - or bidirectional and it would be postulated that initiation of each replicon occurs simultaneously, that each replicon was of a similar size and that the DNA synthesis in each replicon occurred at similar rates. In this way, ery-r2 and tet-A would occur towards the end of one replicon, whilst str-r41 would occur towards the end of another replicon. Evidence of such a mechanism has been put forward by Huberman and Riggs (1968) and by Callan (1972) for the replication of mammalian chromosomes.

The discrepancy between these results and those of

Ravin concerning the linkage of ery-r2 with either str-r41 or str-r53 may be due to differences in the recipient organisms used in the two laboratories. Ravin used strain SIII-I as recipient, which had been derived by Ephrussi-Taylor (1951) by transformation of strain r36A with A66IIIN-DNA, which was followed by at least one mutational event. C13 was isolated by Ephrussi-Taylor (1951) directly from r36A, and hence the two strains have had different histories. Strain C13 has also been used as recipient by Professor Sicard, and he also has never seen evidence of linkage between ery-r2 and str-r41 (personal communication). Ravin, therefore, has kindly presented me with strain SIII-I and DNA prepared in his laboratory from a strain derived from SIII-I but having the ery-r2 and str-r53 markers, so that this system may be compared directly with that of strain C13.

3. STRAIN SIII-I AS RECIPIENT

Making the homologous cross, SIII-I x (SIII-I ery-r2 str-r53), the linkage described by Ravin between ery-r2 and str-r53 has been confirmed. An example of the results is given in Table III, in which the values of the frequencies of the ery-r2 and str-r53 double transformants obtained experimentally are compared with the calculated maxima expected for unlinked markers. The values of the ratio $\frac{\text{reciprocal of experimental frequency}}{\text{reciprocal of calculated frequency}}$ are also given, and these provide a useful assessment of the degree of linkage (and perhaps of non-linkage), of the markers concerned.

However, in the heterologous cross C13 x (SIII-I ery-r2 str-r53) no linkage between ery-r2 and str-r53 was obtained, as seen from results given in Table III, whereas in the other

TABLE III

LINKAGE OF STREPTOMYCIN AND ERYTHROMYCIN RESISTANCE GENES

Expt.	[DNA]	SIII-I x (SIII-I str-r53 ery-r2) str-r53	ery-r2	str-r53 ery-r2 Exptl.	Calc.	Exptl./Calc.	C13 x (SIII-I str-r53 ery-r2) str-r53	ery-r2	str-r53 ery-r2 Exptl.	Calc.	Exptl./Calc.
1	1	93	103	4493	9930	0.45					
2	7	61	61	1941	3758	0.52					
	3 1/2	78	92	3120	7236	0.43					
	1	125	131	8245	16375	0.50					
	1/2	196	173	15635	33908	0.46					
	1/4	333	348	46703	115884	0.40					
	1/10	519	577	56116	299463	0.19					
3	1						94	14	5867	1332	4.40
	1/2						496	61	85073	30306	2.81

Expt.	[DNA]	SIII-I x (C13 - 3Tc3Amlr$_2$S$_{41}$Q)$_{II}$ str-r41	ery-r2	str-r41 ery-r2 Exptl.	Calc.	Exptl./Calc.
4	1	67	77	2490	5170	0.48
5	1	80	88	3712	7040	0.53

heterologous cross SIII-I x $(C13\text{-}3Tc3Am1r_2S_{41}Q)_{II}$, linkage was found again between ery-r2 and str-r41. It is also of interest to note that the frequencies of transformation of each single marker were similar when SIII-I was the recipient, but very different when C13 was the recipient. This is in agreement with the observation of Ravin that there appeared to be no high and low efficiency markers with SIII-I as recipient taking the relative frequencies of transformation as the criterion (personal communication). This, of course, is not the only criterion that should be observed in the determination of high and low efficiency markers.

Further dilution curves were then carried out, using strain SIII-I as recipient to DNA extracted from strains derived from strain C13, in order to investigate possible linkages between combinations of the markers str-r41, ery-r2, opt-r2, amiA-r1 and tet-A. Examples of results obtained are given in Tables IV and V and Fig. 5, and a summary of all results so far obtained is given in Table VI. It can be seen that no linkage occurs between amiA-r1 and ery-r2, tet-A and amiA-r1, and tet-A and opt-r2, but the following pairs are linked: tet-A ery-r2, ery-r2 str-r41, amiA-r1 str-r41, tet-A str-r41, amiA-r1 opt-r2, and ery-r2 opt-r2, the first pair being strongly linked and the last pair being more weakly linked. At the moment, the results are inconclusive regarding the pair str-r41 opt-r2 (Table V). The most concentrated dilution of DNA gave a result indicating no linkage, but the lower concentrations all indicated linkage. This has been noted in the case of tet-A amiA-r1 also, although much greater dilution was needed before the change occurred. The significance of this change is not yet clear, though it may represent the degree of heterogeneity of the DNA solution.

TABLE IV

RECIPROCAL OF THE RATIO $\frac{\text{DOUBLE TRANSFORMANTS}}{\text{C.F.U.}}$ IN THE CROSS SIII-I x $(C13\text{-}3Tc3r_2S_{41}Q)_{II}$

[DNA]	ery-r2 str-r41			tet-A-str-r41			tet-A-ery-r2		
	Exptl.	Calc.	Exptl./Calc.	Exptl.	Calc.	Exptl./Calc.	Exptl.	Calc.	Exptl./Calc.
3	2513	5029	0.50	3825	8378	0.43	130	9049	0.014
1	8932	18514	0.48	9936	21168	0.47	221	22814	0.0097
1/4	26944	56099	0.48	37667	64605	0.58	428	66838	0.0064
1/10	47087	149009	0.32	91476	283237	0.32	1006	284257	0.0035
1/20	91509	521715	0.18	288150	878999	0.33	2058	1035750	0.0020
1/40	308917	2041333	0.15	420657	1738679	0.24	2588	1813266	0.0014
1/100				1746364	11690108	0.15	5368	12593100	0.00042

[DNA]	amiA-r1-ery-r2			amiA-r1-str-r41		
	Exptl.	Calc.	Exptl./Calc.	Exptl.	Calc.	Exptl./Calc.
3	5825	4851	1.21	2965	5056	0.59
1	24103	12884	1.87	8061	13525	0.60
1/4	92321	40938	2.26	26738	43625	0.61
1/10	181222	111609	1.62	92321	97801	0.94
1/20	376385	266506	1.41	212739	290404	0.73

TABLE V

RECIPROCAL OF THE RATIO $\frac{\text{DOUBLE TRANSFORMANTS}}{\text{C.F.U.}}$ IN THE CROSS SIII-Ix $(C13\text{-}3Tc3r_2S_{41}Q)_{II}$

[DNA]	tet-A amiA-r1			str-r41 opt-r2		
	Exptl.	Calc.	Exptl./Calc.	Exptl.	Calc.	Exptl./Calc.
3	18590	9876	1.88	3856	3791	1.02
1	61968	25118	2.47	7788	16308	0.48
1/4	155757	73785	2.11	28929	65360	0.46
1/10	213444	304658	0.70	76861	158417	0.49
1/20	-	-	-	244884	771120	0.32
1/40	480250	1423921	0.34			

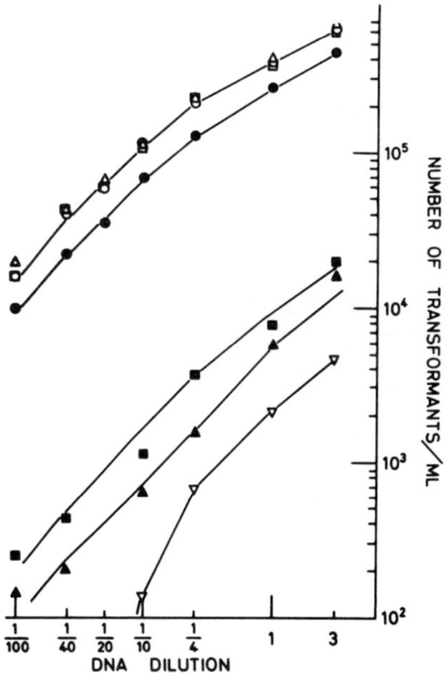

FIG. 5 - Dilution curves for transformants to the markers ery-r2 (O), str-r41 (△), tet-A (□), tet-A ery-r2 (●), ery-r2 str-r41 (■), tet-A str-r41 (▲), and tet-A opt-r2 (▽).

TABLE VI

LINKAGE GROUPS IN THE CROSS SIII-I x $(C13\text{-}3Tc3Am1r_2SQ)_{II}$

		Linked			Unlinked
Strong ⟵			⟶ Weak		
tet-A ery-r2		amiA-r1 str-r41	amiA-r1 opt-r2		amiA-r1 ery-r2
	ery-r2 str-r41				tet-A amiA-r1
		tet-A str-r41	ery-r2 opt-r2		tet-A opt-r2
				str-r41 opt-r2	

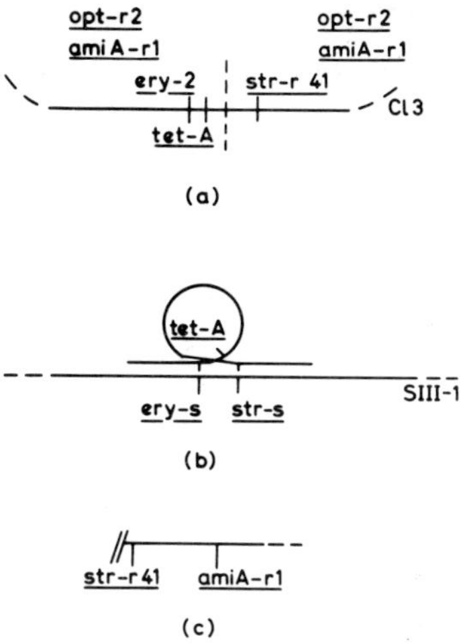

FIG. 6 - Models proposed for the integration of DNA derived from strain C13 by strain SIII-I as recipient. (a) position of markers on the C13 - derived DNA situated towards the end of the two replicons in a chromosome divided into two replicons; the dotted line indicates the end of the two replicons. (b) integration of a loop of DNA possessing the markers ery-r2, tet-A and str-r41 into the SIII-I chromosome; (c) piece of strain C13 - derived DNA possessing the str-r41 and amiA-rl markers; the double lines indicate the point of breakage.

4. DISCUSSION

These results clearly show that there are big differences between strain SIII-I and strain C13, which are presumably due to a different order of the markers on the chromosome of each strain. Until further work has been done, suggestions regarding these differences can only be speculative. Taking as a working hypothesis a model having bi-directional replication from one initiation point (giving two replicons), the distal part of the chromosome of strain C13 would appear as in Fig. 6

(a) with ery-r2 and tet-A towards the end of one replicon, and str-r41 towards the end of the other replicon. It is not known for certain in which replicon the markers opt-r2 and amiA-r1 occur, and hence have been placed in both replicons, but it is known that amiA-r1 replicates before ery-r2 or str-r41, and opt-r2 replicates before amiA-r1. With SIII-I as recipient, ery-r2 and str-r41 appear to be much closer to each other and hence it is possible that when C13-derived DNA is the donor, it is integrated as a loop by pairing at the erythromycin and streptomycin loci, (Fig. 6 (b)). The marker amiA-r1 is not associated with ery-r2 or with tet-A, but it can be associated with str-r41 or opt-r2; it is therefore likely to occur on the same side as str-r41. It is possible therefore that the piece of DNA shown in Fig. 6 (c) may integrate as a separate piece by virtue of a tendency for a breakage to occur to the one side of str-r41. However, the position of opt-r2 is not yet definable; it can be associated with ery-r2 and perhaps with str-r41. There may, of course, be several replicons with or without bi-directional replication, as suggested in Figure 4 (b).

In conclusion, it is clear that more than one replicon exists in the pneumococcus chromosome, and that the order of the markers on the replication map is not identical with the order on the chromosome itself. It is also clear that the order of markers is different on strain SIII-I when compared to that on strain C13.

5. REFERENCES

ANAGNOSTOPOULOS, C. & CRAWFORD, I.P. (1961). Proc. Natl. Acad. Sci. (Wash). 47:378

BUTLER, L.O. & NICHOLAS, G. (1972). Submitted for publication.

BUTLER, L.O. & SMILEY, M. (1972). Submitted for publication.

CAHN, F.H. & FOX, M.S. (1968). J. Bact. 95:867.

CALLAN, H.G. (1972). Proc. R. Soc. Lond. B. 181:19.

CARO, L.G. & BERG, C.M. (1968). Cold Spring Harbor Symposium of Quantitative Biology, 33:559.

CARO, L.G. & BERG, C.M. (1969). J. Mol. Biol., 45:325.

DUBNAU, D., GOLDTHWAITE,C., SMITH, I. & MARMUR, J. (1967).J.Mol. Biol. 27:163.

EPHRUSSI-TAYLOR, H. (1951). Exptl. Cell Res. 2:589.

FOX, M.S. & HOTCHKISS, R.D. (1957). Nature (Lond.) 179:1322.

HUBERMAN, J.A. & RIGGS, A.D. (1968). J. Mol. Biol. 32:327.

MASTERS, M. & BRODA, P. (1971). Nature, New Biology, 232:137.

NESTER, E.W. & STOCKER, B.A.D. (1963). J. Bact. 86:785.

NISHIOKA, Y. & EISENSTARK,A. (1970). J. Bact. 102:320.

O'SULLIVAN, A. & SUEOKA, N. (1967). J. Mol. Biol. 27:349.

PORTER, R.D. & GUILD, W.R. (1969). J. Bact. 97:1033.

RAVIN, A.W. (1966). Symposium on the Mutational Process. The Physiology of the Gene and Mutation Expression.Czeck. Acad. Sc. p. 139.

RAVIN, A.W. & CHEN, K.C. (1967). Genetics, 57:851.

YAHARA, I. (1971). J. Mol. Biol. 57:373.

TRANSFORMATION IN THE BLUE-GREEN ALGA *ANAC*
AND AN ASSOCIATED PHENOMENA OF MUT

MICHAEL HERDMAN

Department of Biochemistry and Department of Genetics,
The University of Liverpool, Liverpool L69 3BX

ABSTRACT

Two systems of transformation operate in the procaryotic alga *Anacystis nidulans*. One, mediated by chemically extracted DNA, resembles bacterial transformation in being sensitive to DNase. The second is mediated by an extracellular DNA/RNA complex secreted into the growth medium, and is sensitive to both DNase and RNase although neither enzyme totally destroys the transforming activity of this material. Linkage is more extensive in transformation mediated by extracellular nucleic acids than in transformation with chemically extracted DNA. A high proportion of transformants in both systems are found to be mutant in loci which were wild type in both parental strains. The frequency of mutation is related to the frequency of recombination, and the mutations appear to arise at, or near to, the sites of recombination. Evidence suggests that the recombinational event itself is mutagenic in transformation in *A. nidulans*.

1. INTRODUCTION

The blue-green algae are procaryotic organisms which, in all important points of comparison, may be grouped closely with the bacteria. In contrast to the rapid progress that has been made in our understanding of bacterial genetics, progress in blue-green algal genetics has been surprisingly slow. Early problems encountered in quantitative surface growth were overcome by Van Baalen (1965) and Allen (1968) and difficulties in the isolation of auxotrophic mutants have been partially overcome (Herdman & Carr, 1972).

Recombination occurs at low frequency in mixed cultures of mutant strains of <u>Anacystis nidulans</u> (Bazin, 1968). Genetic transformation by <u>chemically extracted DNA</u> has been described in <u>A. nidulans</u> (Shestakov & Khyen, 1970) and in the same organism a second, unusual, type of transformation mediated by <u>extracellular nucleic acids</u> secreted into the growth medium also occurs (Herdman & Carr, 1971). The use of these processes of gene transfer in genetic mapping has not been reported, although temporal genetic mapping by sequential mutagenesis during synchronous growth has been successfully employed (Asato & Folsome, 1970; Herdman, Faulkner & Carr, 1970).

This communication describes some of the characteristics of transformation mediated by both chemically extracted DNA and extracellular nucleic acids, and presents evidence that in both systems genetic recombination is associated at high frequency with recombination.

2. METHODS

<u>Organism</u>

<u>Anacystis nidulans</u> (No. 625 of the Culture Collection of Algae at Indiana University, Department of Botany, Bloomington, Indiana, USA) was cultured as described (Herdman, Faulkner & Carr,

1970). This organism was employed in these studies because it is unicellular, grows rapidly, forms homogeneous suspension in liquid culture, and is easily cultured on the surface of agar. The mutants employed are listed in Table 1 and were isolated and cultured as described (Herdman & Carr, 1972).

TABLE I

MUTANTS OF *ANACYSTIS NIDULANS*

Strain	Markers
1111	*nit-1*
1121	*cys-1*
1141	*bio-1*
2111	*nit-1 str-1*
2113	*nit-1 met-2*
2121	*cys-1 str-3*
2141	*bio-1 str-2*
3141	*bio-1 nit-2 str-2*

Transformation procedure

Preparation of donor material. Transforming DNA was extracted as described by Craig, Leach & Carr (1969), except that extrusion through the French Pressure Cell was performed at no more than 1000 lb/sq in pressure. Donor cell-free filtrates were prepared by Millipore filtration through a 0.45 μ filter of cultures (10 or 20 ml) containing 5 x 10^8 cells/ml. Such cell-free filtrates were used as the normal source of donor extracellular nucleic acids, except in some experiments where extracellular nucleic acids were isolated from culture medium,after harvesting the cells, by ultrafiltration at 0-4° in a 2 l ultrafiltration cell with an Amican UM-10 150 mm membrane (Chemlab Instruments Ltd.,Ilford, Essex); nucleic acids were then extracted from this concentrate and purified by the normal DNA purification procedure (Craig, Leach & Carr, 1969).

Incubation conditions. Recipient organisms were harvested during exponential growth and resuspended to 5x10^7 cells/ml in growth medium (20 ml). Donor chemically extracted DNA (0.1 µg/ml) was added to the recipient culture and incubated at 34° 4 in from three warm white fluorescent lights(40 W)and gassed with air:CO_2 (95:5,v/v).Following incubation with DNA,DNase(50 µg/ml) plus $MgCl_2$ (5 µmoles/ml) were added and incubated at 34° for 20 min to degrade remaining donor DNA. Donor cell-free filtrates (2 ml) were incubated with recipient culture as described above, except that DNase was not normally added since this enzyme does not totally degrade the donor material (Herdman & Carr, 1971). Where purified extracellular nucleic acids were used, the sample was first diluted to a concentration equivalent to that of the extracellular nucleic acids in normal cell-free filtrates and then used in transformation as described above. The effect of DNase, RNase and pronase on transforming activity in certain experiments was examined by pre-incubation with enzyme (50 µg/ml) for 20 min at 34° . DNase was added in the presence of $MgCl_2$ as described above.RNase was prepared in 0.15 M NaCl and pretreated at 80° for 10 min to destroy any traces of DNase. Pronase was preincubated at 34° for 15 min to destroy any traces of both nucleases.Ultrasonic treatment of cell-free filtrates in certain experiments was carried out in an M.S.E. 60 W, 20 Kc/sec sonic disintegrator at maximum output for 10 sec.

Selection of transformants and screening for mutations. Transformants were scored by plating aliquots (0.1 ml) of the recipient culture containing 2.5×10^8 cells/ml on normal selective media.Mutations arising during the course of the experiment were isolated by plating recipient cultures on selective media which contained supplements additional to those required by the parental strains. The supplements used enabled mutations conferring the Bio, Cys, Met and Nit phenotypes to be isolated. Individual mutations were subsequently identified by replica plating. Supplements were employed at the concentrations previously described (Herdman & Carr, 1972).The visible mutation Fil, conferring filamentous growth, was also recorded.

3. RESULTS AND DISCUSSION

Transformation by extracellular nucleic acids was first observed in control experiments designed to investigate recombination in mixed cultures of mutant strains of *A. nidulans*, using the classical U-tube technique (Davis, 1950) in which two mutant strains were grown in the two arms of the U-tube separated by a sinter which was impermeable to cells but permitted the flow of growth medium. Unexpectedly, recombination occurred at high frequency in the absence of contact between the cells of the two strains, and therefore a process of genetic transfer,mediated by cell-free culture filtrate, had operated. In subsequent experiments cell-free filtrates were prepared by filtration, as described in Methods,and used as the source of donor genetic material usually without further treatment.

The outstanding features of transformation mediated by extracellular nucleic acids were that only a few of the possible recombinant classes were recovered, and that the donor parental class occurred at highest frequency. These features are illustrated in the two experiments shown in Table II, and in Experiment 2 especially the donor class occurred at extremely high frequency

TABLE II

TRANSFORMATION BY EXTRACELLULAR NUCLEIC ACIDS

Exp. No.	Parental Strains	Recombinant classes	Frequency ($\times 10^{-6}$)
1 D	*nit-1*$^+$ *met-2*$^+$ *cys-1* *str-3*	*nit-1*$^+$ *met-2*$^+$ *cys-1* *str-3*	15.20
R	*nit-1* *met-2* *cys-1*$^+$ *str-3*$^+$	*nit-1*$^+$ *met-2*$^+$ *cys-1* *str-3*$^+$	0.11
		nit-1$^+$ *met-2* *cys-1* *str-3*$^+$	0.31
		nit-1$^+$ *met-2* *cys-1*$^+$ *str-3*$^+$	1.15
2 D	*bio-1* *cys-1*$^+$ *str-2*	*bio-1* *cys-1*$^+$ *str-2*	1200
R	*bio-1*$^+$ *cys-1* *str-2*$^+$	*bio-1*$^+$ *cys-1*$^+$ *str-2*	0.08
		bio-1$^+$ *cys-1*$^+$ *str-2*$^+$	0.03

Donor material was freshly prepared cell-free filtrate.
D = donor, R = recipient.

in relation to the other classes. A wide range of strains of *A. nidulans* were examined, and all those tested were capable of acting as both donors and recipients in this system of transformation.

The effects of DNase,RNase and the general proteolytic enzyme pronase on the formation of recombinants are shown in Table III. Pre-treatment of donor cell-free filtrate with DNase

TABLE III

THE EFFECTS OF PRETREATMENT WITH DNase,RNase AND PRONASE ON TRANSFORMATION WITH EXTRACELLULAR NUCLEIC ACIDS

Enzyme	Frequency of transformants ($\times 10^{-5}$) *
-	1.20
DNase	0.48
RNase	0.49
DNase + RNase	0.48
Pronase	1.22

* Transformants inheriting the donor genotype in the cross of 1121 cells with 2141 filtrate.

reduced the frequency of transformants from 1.2×10^{-5} to 4.8×10^{-6}, or by only 60%.Unexpectedly,pretreatment with RNase had a similar effect. Treatment with both DNase and RNase together did not further reduce the number of recombinants, and pronase had no effect. The action of DNase on the transforming activity of the preparation supported the view that gene transfer was mediated by a process of transformation, since this action of DNase is observed in bacterial transformation. The effect of RNase, however, was unusual. In contrast, pretreatment of the DNA with RNase had no significant effect on transformation mediated by chemically extracted DNA, whilst DNase completely removed the transforming activity of such preparations (Table IV). The two processes

TABLE IV

THE EFFECTS OF PRETREATMENT WITH DNase AND RNase ON TRANSFORMATION WITH CHEMICALLY EXTRACTED DNA

Enzyme	Frequency of transformants (x 10^{-5})
-	4.8
DNase	0
RNase	4.7

Recipient strain 3141, prototrophic DNA.

of transformation therefore differed in their responses to nucleases, and extracellular nucleic acids were extremely unusual in their sensitivity to RNase and in their partial resistance to DNase. It has previously been suggested (Herdman & Carr, 1971) that both DNA and RNA are necessary for efficient transformation in this system, and that the transforming principle may be a DNA/RNA complex. A nucleic acid complex of an apparently similar nature has been observed in Pneumococcus (Evans, 1964) where transformation mediated by RNA fractions extracted from donor cells was sensitive to both DNase and RNase. In this case, however, the transformants initially behaved as unstable heterozygotes in which the transferred characters were either ultimately lost or stably inherited. In *A. nidulans* the transformants appear to be stable, since no evidence of segregation of markers has been obtained.

Both DNA and RNA were isolated from cell-free culture medium of *A. nidulans* by the normal purification procedure following concentration by ultrafiltration. This nucleic acid preparation was active in transformation when diluted back to the initial concentration (Herdman & Carr, 1971), and was sensitive to both DNase and RNase to the same extent as normal cell-free filtrates. The relative ease of isolation of the extracellular nucleic acids in an apparently undamaged form will be of great value in the examination of their physical properties.

Genetic mapping experiments using both systems of transformation produced inconsistent linkage values,for reasons which will be described below. For example, cotransfer indices ranging from 0.4 to 0.9 were recorded for the markers *bio-1* and *str-2*, and from 0.3 to 0.99 for *nit-1* and *str-2*, in transformation with extracellular nucleic acids. Although map distances cannot, at present, be assigned to the markers, the results obtained were consistent with the map order for five markers shown in Fig. 1.

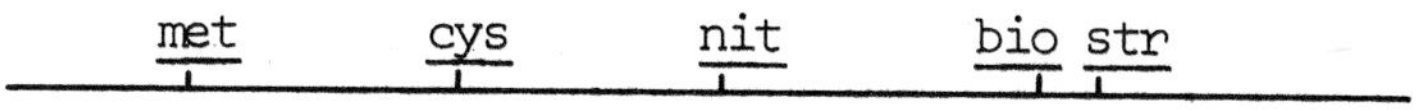

Fig. 1 - MAP ORDER FOR FIVE MARKERS OF *A. NIDULANS*

The gene order was deduced from experiments with extracellular nucleic acids. Map distances are not included due to the inconsistencies in linkage values described in the text.Markers *bio* and *str* are shown close together because these were the only markers to be cotransformed by chemically extracted DNA.

For comparison, the only other available genetic map of A. nidulans, produced by sequential mutagenesis (Asato & Folsome, 1970), is shown in Fig. 2. This map shows the positions of six markers on

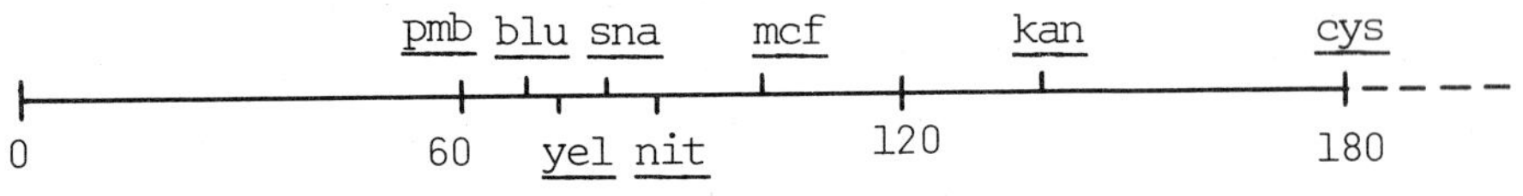

Fig. 2 - GENETIC MAP OF A. NIDULANS PRODUCED BY SEQUENTIAL MUTAGENESIS (AFTER ASATO & FOLSOME, 1970)

The time scale is shown in minutes and the full scale represents 180 minutes of the chromosome replication time of 420 minutes. The markers nit-1 and cys-1 have been placed on the same scale as described in the text.

a 420 minute scale representing the chromosome replication time. In our system of synchronous growth (Herdman, Faulkner & Carr, 1970) the genome replication time was 60 minutes, and the nit-1 and cys-1 markers replicated 12 and 26 minutes respectively after initiation of replication. By adjustment of this time scale to the 420 minute map of Asato & Folsome (1970) nit-1 and cys-1 lie at 84 and 182 minutes respectively on this map and therefore the regions of the genome shown in Figs. 1 and 2 overlap.

The linkage of markers in transformation mediated by extracellular nucleic acids was more extensive than in that mediated by chemically extracted DNA. For example, in transformation with chemically extracted DNA up to 57% of the transformants inherited both bio-1 and str-2, whilst these markers were cotransformed at higher frequencies of up to 96% in transformation with extracellular nucleic acids. More widely spaced markers, for example *met-1* and *nit-1*,were not cotransformed by chemically extracted DNA, but up to 79% of transformants produced by extracellular nucleic acids inherited the entire *met—str* region. It is therefore clear that, in comparison with chemically extracted DNA, large segments of donor genome are transformed intact in transformation with extracellular nucleic acids.

Extracellular DNA is also active in transformation in *Neisseria meningitidis* (Catlin,1960), *Pneumococcus* (Ottolenghi & Hotchkiss, 1960) and *Bacillus subtilis* (Takahashi, 1962). Increased linkage was observed in transformation with this DNA in *B. subtilis* (Ephrati-Elizur, 1968) when compared to the normal extracted DNA, and in this respect this organism compares with *Anacystis nidulans*.

When additional supplements were employed in the selective medium to permit the growth of phenotypes other than those which were expected to arise by recombination, mutations were observed in loci which were wild type in both parental strains. Such mutations did not arise in control experiments in which recipient cells were plated without contact with donor material, or where donor material alone was plated. Table V shows the mutations produced in a cross between strain 2141 and prototrophic donor material in the two systems of transformation. With chemically extracted DNA (Experiment 1) only one transformant class was produced, inheriting the donor *bio-1*$^{+}$ marker. A small number of these transformants carried no detectable mutations, but the majority carried mutations in either the nitrate reductase (*nit*) or sulphate reduction (*cys*)genes and occurred at appreciable frequencies. With extracellular nucleic acids (Experiment 2) again only one transformant class was produced,but this inherited both

TABLE V

MUTATIONS ARISING IN THE CROSS OF STRAIN 2141 (bio-1 str-2) WITH PROTOTROPHIC DNA AND EXTRACELLULAR NUCLEIC ACIDS

Exp. No.	Transformant Class	Mutant Phenotype	Frequency (x 10^{-6})
1	$bio\text{-}1^{+}$ str-2	-	0.04
		Nit	5.2
		Cys	2.5
2	$bio\text{-}1^{+}$ $str\text{-}2^{+}$	-	71
		Nit	53

The donor material in Experiment 1 was chemically extracted DNA, and in Experiment 2 was extracellular nucleic acids in cell-free filtrate.

donor markers. A high proportion of these transformants carried no detectable mutations, and a slightly lower proportion were mutant in the nitrate reductase locus. A series of similar experiments showed that spontaneous mutation occurred in all experiments with chemically extracted DNA and in a high proportion (70%) of experiments with extracellular nucleic acids. In addition, it became clear that the positions of the mutagenic events were associated with the recombinational events. For example, the cross with chemically extracted DNA shown in Table VI produced two progeny classes: the first inherited only str-3 by recombination in the region between bio-1 and str-3 (see Fig. 1), and did not give rise to mutations; the second class was produced by recom-

TABLE VI

MUTATIONS ARISING IN THE CROSS OF STRAIN 1141 (bio-1) WITH DNA EXTRACTED FROM STRAIN 1121 (cys-1 str-3)

Transformant Class	Mutant Phenotype	Frequency ($\times 10^{-6}$)
cys-1$^+$ bio-1 str-3	-	0.01
cys-1$^+$ bio-1$^+$ str-3	Nit	1.0
	Nit Fil	0.79

bination between cys-1 and bio-1, inherited both bio-1$^+$ and str-3, and all carried mutations in the nitrate reductase locus. In addition, some were mutated in a second gene, fil, which gave rise to a filamentous phenotype. Since nit lies between cys-1 and bio-1 (Fig. 1), and recombination occurred in this region, it may be suggested that nit mutation arose as a consequence of recombination within the nit locus. The Fil mutation may represent recombination and mutation at the other end of the transferred DNA, and would therefore place fil to the right of str on the genetic map. Similarly, in the cross of Table VII (extracellular nucleic acids) mutations arose in the sulphate reduction gene (cys) only in transformants produced by recombination between met-2 and bio-1, i.e. in the region of the cys locus which became mutant. These and further experiments suggested that the positions of the mutational and recombinational events were closely related. Evidence in support of this hypothesis was obtained in crosses with extracellular nucleic acids when the molecular weight of the donor

TABLE VII

MUTATIONS ARISING IN THE CROSS OF STRAIN 2113 (nit-1 met-2) WITH EXTRACELLULAR NUCLEIC ACIDS IN CELL-FREE FILTRATE OF STRAIN 2141 (bio 1 str-2)

Transformant Class	Mutant Phenotype	Frequency (x 10^{-7})
$met\text{-}2^+$ $nit\text{-}1^+$ bio-1 str-2	-	39
$met\text{-}2^+$ $nit\text{-}1^+$ bio-1 $str\text{-}2^+$	-	0.23
$met\text{-}2^+$ nit-1 bio-1 $str\text{-}2^+$	-	1.4
$met\text{-}2^+$ nit-1 $bio\text{-}1^+$ $str\text{-}2^+$	Cys	4.6

material was reduced by sonication. In the normal cross only one transformant class was produced, inheriting the entire cys-1—str-2 region and none of these transformants exhibited new mutations; following sonication of the same donor material recombination was observed between cys and bio in every transformant, and all these transformants carried Nit mutations (Table VIII). Once again, the positions of the mutational and recombinational events were related, since recombination in the region of the nit locus produced Nit mutants, and mutation occurred only with concommitant recombination. The reduction in linkage in this experiment resulting from sonication of the cell-free filtrate caused this material to behave like chemically extracted DNA which, in the identical cross (Table VI) produced Nit and Nit Fil mutants in similar proportions.

TABLE VIII

RECOMBINATION AND MUTATION FOLLOWING SONICATION OF DONOR EXTRACELLULAR NUCLEIC ACIDS

Treatment	Transformant Class	Mutant Phenotype	Frequency
Normal cross	$cys\text{-}1^{+}$ bio-1 str-2	-	2.04×10^{-5}
Sonication	cys-1 bio-1 str-2	Nit	2.1×10^{-8}
		Nit Fil	2.0×10^{-8}

Parental strains: donor $cys\text{-}1^{+}$ bio-1 str-2

recipient cys-1 $bio\text{-}1^{+}$ $str\text{-}2^{+}$

It is clear that the occurrence of mutation during recombination will affect the accuracy of genetic mapping. In the absence of a complete medium which would support the growth of all transformants in which mutation occurs, only a narrow range of phenotypes may be selected. Many unselected mutations will therefore be lethal, and the transformants in which they occur will be lost. In addition, the distortion of linkage values resulting from incorrect classification of transformants following mutation of the parental markers themselves cannot be excluded. It may be suggested that the mutational phenomenon is responsible for the inconsistent linkage values described earlier.

Spontaneous mutation has been described in the meiotic or recombinational cycles of a variety of microorganisms, including *Salmonella typhimurium* (Demerec, 1962), *Bacillus subtilis*

(Yoshikawa, 1966) and *Saccharomyces cerevisiae* (Magni, 1963), and appeared to occur at the sites of recombination as shown by the recombination of outside markers. In all cases, however, the mutation frequencies were low, representing approximately a 10-fold increase over the mitotic or non-recombinational rates. In *A. nidulans* the frequency of mutation associated with recombination is extremely high and may represent a fundamental error in the recombinational process in this organism.

In this discussion only a few examples have been used to illustrate the general characteristics of mutation associated with recombination; the experimental results have been described in greater detail elsewhere (Herdman, in press). The supplements employed in these experiments have permitted the isolation of only a narrow range of mutants. A defined complete medium for *A. nidulans* has recently been developed (Herdman, Delaney & Carr, unpublished) which will permit a more complete analysis of the mutational phenomenon. The occurrence of this phenomenon, and of transformation mediated by extracellular nucleic acids, in the natural environmant remain to be investigated.

ACKNOWLEDGEMENTS

I wish to thank Dr. N. G. Carr and Dr. B. M. Faulkner for advice and guidance in this work and I am grateful for support from the Science Research Council.

4. REFERENCES

ALLEN, M. M. (1968). J. Phycol. 4:1 - Simple conditions for the growth of unicellular blue-green algae on plates.

ASATO, Y. & FOLSOME, C. E. (1970). Genetics 65:407 - Temporal genetic mapping of the blue-green alga *Anacystis nidulans*.

BAZIN, M. J. (1968). Nature (London) 218:282 - Sexuality in a blue-green alga:genetic recombination in *Anacystis nidulans*.

CATLIN, B. W. (1960). J. Bact. 79:579 - Transformation of *Neisseria meningitidis* by deoxyribonucleates from cells and from culture slime.

CRAIG, I. W., LEACH, C. K. & CARR, N. G. (1969). Arch. Mikrobiol. 65:218 - Studies with deoxyribonucleic acid from blue-green algae.

DAVIS, B. D. (1950). J. Bact. 60:507 - Non-filterability of the agents of genetic recombination in *E. coli*.

DEMEREC, M. (1962). Proc. Natl. Acad. Sci. (Wash.) 48:1696 - "Selfers"-attributed to unequal crossovers in *Salmonella*.

EPHRATI-ELIZUR, E. (1968). Genet. Res. 11:83 - Spontaneous transformation in *Bacillus subtilis*.

EVANS, A. H. (1964). Proc. Natl. Acad. Sci. (Wash.) 52:1442 - Introduction of specific drug-resistance properties by purified RNA-containing fractions from *Pneumococcus*.

HERDMAN, M. & CARR, N. G. (1971). J. Gen. Microbiol. 68:xiv - Recombination in *Anacystis nidulans* mediated by an extracellular DNA/RNA complex.

HERDMAN, M. & CARR, N. G. (1972). J. Gen. Microbiol. 70:213 - The isolation and characterization of mutant strains of the blue-green alga *Anacystis nidulans*.

HERDMAN, M., FAULKNER, B. M. & CARR, N. G. (1970). Arch. Mikrobiol. 73:238 - Synchronous growth and genome replication in the blue-green alga *Anacystis nidulans*.

MAGNI, G. E. (1963). Proc. Natl. Acad. Sci. (Wash.) 50:975 - The origin of spontaneous mutations during meiosis.

OTTOLENGHI, E.& HOTCHKISS, R. D. (1960). Science 132:1257 - Appearance of transforming activity in *Pneumococcal* cultures.

SHESTAKOV, S. V. & KHYEN, N. T. (1970). Mol. Gen. Genetics 107: 372 - Evidence for genetic transformation in the blue-green alga *Anacystis nidulans*.

TAKAHASHI, I. (1962). Biochem. Biophys. Res. Commun. 7:467 - Genetic transformation of *Bacillus subtilis* by extra-cellular DNA.

YOSHIKAWA, H. (1966). Genetics 54:1201 - Mutations resulting from the transformation of *Bacillus subtilis*.

VAN BAALEN, C. (1965). J. Phycol. 1:19 - Quantitative surface plating of coccoid blue-green algae.

ANALYSIS OF STEPS IN PENETRATION AND INTEGRATION OF BACTERIAL DNA IN NORMAL OR X-IRRADIATED MICE FIBROBLASTS

FRANÇOISE LAVAL
EDMOND MALAISE

Groupe de Recherche de Radiobiologie Clinique (INSERM)
Institut Gustave-Roussy, 94 Villejuif, France

ABSTRACT

As judged by CsCl gradient centrifugation analysis, *E. coli* DNA is taken up by NCTC cells as a double-stranded molecule. This bacterial DNA is then integrated in the DNA of the recipient cells without being degraded. Pieces of bacterial DNA are covalently linked to each of the two strands of the recipient DNA during the cellular DNA synthesis. The repair synthesis following large doses of X-irradiation is not implicated in this integration process.

1. INTRODUCTION

Many reports have appeared on the uptake of homologous and heterologous DNA by various mammalian cells, either *in vitro* or *in vivo*. They have been extensively reviewed by BHARGAVA & SHANMUGAM(1971). These reports deal with the uptake and biological effects of foreign DNA. However, the fate of the exogenous DNA in the recipient cells was not well established.

In order to investigate the fate of heterologous DNA in mammalian cells in vitro, we have choosen to use labeled E.coli DNA, which could be detected by its radioactivity, characterized and separated by its buoyant density in CsCl gradients from the cellular DNA.

We also tried to correlate the fate of foreign DNA to the cell cycle, and experiments described in this paper were undertaken to examine possible involvement of the cellular DNA synthesis and/or repair synthesis in the integration of heterologous DNA in mammalian cells.

2. MATERIALS AND METHODS

(a) Materials

[^3H]-thymidine, (20 Ci/m Mole) and [2 ^{14}C] thymine (51 m Ci/m Mole) were purchased from CEA Saclay, France - Pancreatic DNase and RNase were obtained from Worthington Biochemical Corporation.

(b) Cells

NCTC cells, originally isolated from C3H mice fibroblasts (SANFORD, LIKELY & EARLE 1954), were obtained from the American Type Culture Collection. They were grown in monolayers, on NCTC 109 medium supplemented with 10% horse serum, and were used during exponential growth. Cells cultures were free of Mycoplasma contamination.

(c) Preparation of DNA

Labeled E. coli DNA was used for the present study - Bacteria were labeled with $[2-^{14}C]$ thymine (1 m Ci/liter), then lysed in SSC (0,15 M NaCl, 0,015 M Na citrate).DNA was extracted by adding distillated phenol (pH 9,0) and shaking for 15 min. Phases were separated and DNA reextracted by phenol. After dial-lysis, RNA was removed by digestion by RNase, and a third phenol treatment was then performed. DNA was precipitated with ethanol, and redissolved in SSC. Any residual phenol was removed by dia-lysis.Specific activities of $[^{14}C]$ - DNA preparations ranged from 50.000 to 70.000 cts/min/µg DNA.

(d) Uptake experiments

Monolayers of NCTC cells (10^7 to $1,5 \times 10^7$ cells) were incubated in Hanks solution, and $[^{14}C]$ E. coli DNA was added to give a final concentration of 1 µg/ml. After incubation at 37°C, uptake was stopped by the addition of pancreatic DNase (2 µg/ml) 15 min at 37° C. After washing the monolayers two times with saline solution, cells were harvested by trypsination. Following trypsination,all experiments were carried out at 4° C.Cells were washed by centrifugation two times with Hanks solution,the final pellet was resuspended in Tris 0,01 M pH 8,2 and cells lysed by sodium dodecyl sulfate (0,1% final concentration).

The total cell mass of DNA was extracted by the procedure described by SCHMIDT & TANNHAUSER (1945). Cellular DNA content was determined by the colorimetric method described by BURTON (1956), and the labeled DNA taken up by the cells was estimated by radioactivity counting of the hydrolysates.

(e) CsCl gradient centrifugation

Buoyant density determinations were performed in a fixed angle n° 40 rotor or in a SW 65 rotor of a Spinco L265K. CsCl was added to cell lysates and density was adjusted by refractometry to a mean value of 1,700 g/cm^3.

After centrifugation (34,000 rev/min, 60h, 15° C for rotor 40; 35,000 rev/min, 48h, 15° C for rotor SW 65), the tubes were punctured and fractions of three drops collected in 1 ml of SSC.

Optical densities at 260 nm were measured with a Zeiss spectrophotometer.

For alkaline CsCl density gradient centrifugation, the pH was adjusted to pH 13,0 with 1N - NaOH. An increment of 0,062 g/cm^3 was allowed between the density of native DNA in neutral ClCs and that of denatured DNA at high pH (VINOGRAD, MORRIS, DAVIDSON & DOVE, 1963).

(f) DNA - DNA hybridization

DNA - DNA hybridization were carried out on Millipore filters according to the technique described by DENHARDT (1966). DNA was denatured by heating at 100° for 10 min followed by chilling in an ice bath.

(g) Radioactivity counting

0,5 ml of fractions from CsCl gradients were dissolved in 5 ml Instagel(Packard)and counted in a Packard Tricarb liquid-scintillation spectrometer model 3375.

3. RESULTS

(a) Uptake of E. coli DNA

The uptake of $[^{14}C]$ - E. coli DNA was estimated by following the appearance of radioactivity in the recipient cells. This study describes the influence of different factors on the uptake of bacterial DNA by NCTC cells. It does not account for the physical state of the incorporated DNA.

A maximal uptake of bacterial DNA was observed when NCTC cells were incubated for 60 min, at 37° C, in a saline solution,(Hanks medium),containing 1 µg/ml of bacterial DNA. (LAVAL, MALAISE, LAVAL 1970).

All the radioactivity taken up was incorporated in the nucleus of the cells and was on an acido-precipitable form. It represents the uptake of exogenous DNA undegraded during the incubation period,as checked by zone centrifugation of labeled DNA samples removed from the medium at different times of incubation. (LAVAL, MALAISE, LAVAL 1971).

The uptake was temperature dependent, and was respectively 50% at 18°C and 7% at 4°C of that observed at 37° C. The temperature dependence and the decrease of uptake by metabolic inhibitors(LAVAL, MALAISE & LAVAL 1970) suggested that it was an active transport through the cell membrane.

The molecular weight of exogenous DNA was a critical factor, since a decrease of uptake of 50% was observed when the size of DNA molecules was reduced from 20S to 11S by sonication.

Denatured DNA, prepared by heating or alkali denaturation, was not accomodated by the uptake system, in conditions consistent with the cell viability.

The radioactivity measured in the recipient cells, represented a maximal uptake of about $1,5 \times 10^{-2}$ μg of bacterial DNA for 100 μg of cellular DNA.This value was independent of the origin of bacterial DNA,as similar results were obtained with T7 or Micrococcus radiodurans DNA.It only accounted for the size of the foreign DNA molecules.

(b) Fate of E. coli DNA

The fate of E. coli DNA after its penetration into the cells was followed by buoyant CsCl density gradient centrifugation and molecular hybridization.Using these techniques, one can answer the following questions:after entering the cells, what is the buoyant density of the foreign DNA? Does it penetrate as a double-stranded or as a single-stranded molecule? Is the foreign DNA integrated in the cellular genome, and if so, is it as a native molecule or as degradation products ? Is the exogenous DNA covalently linked to cellular DNA?Answers to these questions are given by the following experiments:

Neutral ClCs gradient centrifugation

NCTC cells were lysed after different times of incubation with $[^{14}C]$-E. coli DNA.The cell lysates were then fractionated by CsCl density gradient centrifugation. (Fig. 1)

The radioactivity first appeared at the density of native E. coli DNA, (ρ = 1.700 g/cm^3) for cells incubated for 0 to

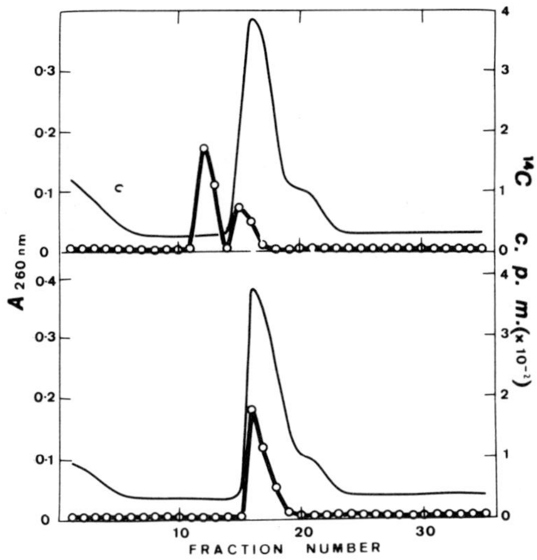

Fig. 1 - Neutral CsCl density gradient of NCTC cells lysed after a 60 min or 120 min incubation period with [^{14}C] - E. coli DNA.—— absorbance; 0———0 [^{14}C] cts/min.

30 minutes with [^{14}C] - E. coli DNA,and then appeared in the cellular DNA (ρ = 1.700 g/cm^3). In the gradients from the 60 min cells,66,6% of the radioactivity taken up by the cells was found at the buoyant density of native E. coli DNA,and 33% banded with the cellular DNA. After a period of 120 min after the beginning of the uptake,all the radioactivity had left the density of bacterial DNA and appeared in the peak of cellular DNA. The distribution of radioactivity, in terms of incubation time, among the densities of bacterial and cellular DNA is represented in TableI.

Results are expressed in per cent of the maximum radioactivity measured in NCTC DNA at 120 min of incubation.

TABLE I

PARTITION OF INCORPORATED RADIOACTIVITY AMONG *E.COLI* AND NCTC DNA BUOYANT DENSITY FRACTIONS IN CsCl GRADIENTS PERFORMED AFTER DIFFERENT TIMES OF INCUBATION BETWEEN NCTC CELLS AND $[^{14}C]$ - *E. COLI* DNA

Incubation time in minutes	% of total $[^{14}C]$ incorporated	
	E.coli DNA density	NCTC DNA density
0	0	0
15	48	0
30	80	0
45	84	11
60	66	34
90	29	71
120	0	100

No radioactivity was detected at the position of denatured *E. coli* DNA, and similar gradients patterns were obtained with deproteinized cells lysates or with isolated nuclei of NCTC cells.

Thus, the exogenous DNA determined in the recipient cells retained its characteristic buoyant density during the first 60 min of uptake: the *E. coli* DNA was taken up as a double stranded molecule. The observed transfer of radioactivity from the density of *E.coli* DNA to that of cellular DNA may be explained in two ways : the foreign DNA can be integrated as a native molecule in the cellular genome,or it can be degraded by cellular nucleases and used as nucleotides by the recipient cells.

Recentrifugation experiments

In the cells incubated for 120 min, the radioactivity was not evenly distributed in the peak of cellular DNA, but was localized in the heavier part of this peak. This could suggest that bacterial DNA was linked to some molecules of cellular DNA, and in this case it should be possible to separate these radioactive molecules from the bulk of cellular DNA by recentrifugation. The results of this experiment are represented on Fig. 2 : After 120 min incubation with [^{14}C] - E. coli DNA, the cells lysate was centrifuged in neutral CsCl. [Fig. 2(a)] ,and radioactive fractions were then pooled and recentrifuged in the same conditions [Fig. 2(b)]. The radioactive fractions were recentrifuged a third time in the same conditions [Fig. 2(c)]. After the third centrifugation two peaks were resolved:one corresponding to cellular DNA (ρ = 1.700 g/cm^3), the other to the labeled DNA integrated into the cells (ρ = 1.704 g/cm^3).

In control experiments, the labeled exogenous DNA was replaced by a corresponding amount of [3H] -thymidine. These experiments showed that,in this case,radioactivity taken up by the cells could not be separated from the cellular DNA (Fig. 2c,d,e).

Therefore the appearance of the radioactivity at a density of 1.704 g/cm^3 was not due to some artefact during the recentrifugation experiments.It rather suggests that bacterial DNA taken up by the NCTC cells was associated with cellular DNA molecules sufficiently large to obscure its characteristic buoyant density.

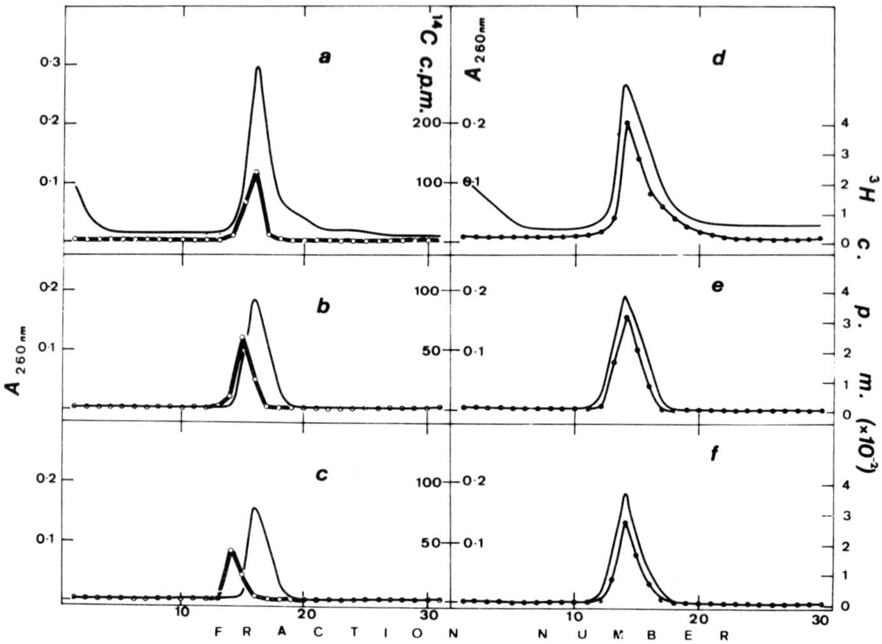

Fig. 2 - Recentrifugation experiment: (a) neutral CsCl density gradient of NCTC cells lysed after 120 min of incubation with [^{14}C] - <u>E. coli</u> DNA; (b) recentrifugation of radioactive fractions under brackets from (a); (c) recentrifugation of radioactive fractions under brackets from (b); (d, e, f) same experiment, but cells are incubated for 120 min with [^{3}H]-thymidine. ——— absorbance; 0——0 [^{14}C] cts/min; ●——● [^{3}H] cts/min.

Hybridization experiments

Experiments of DNA - DNA hybridization were run according to the procedure described by DENHARDT (1966). The results are summarized in Table II. <u>E. coli</u> DNA sequences were found in the

TABLE II

HYBRIDIZATION EXPERIMENTS OF E. COLI DNA AND DNA FROM NCTC CELLS EXPOSED OR NOT TO $[^{14}C]$ - E. COLI DNA

DNA on Millipore filters	DNA hybridized	% of hybridization
E. coli	*E. coli*	52
	DNA of NCTC cells labeled with $[^{3}H]$ - thymidine	3
	DNA of NCTC cells exposed to $[^{14}C]$ - *E. coli* DNA	42

DNA extracted from NCTC cells after incubation with $[^{14}C]$-*E.coli* DNA for 120 min.

Hybridization and recentrifugation experiments strongly suggest that bacterial DNA was integrated in NCTC DNA without being degraded.

Alkaline ClCs centrifugation

Cell lysates from 120 min incubation were subjected to centrifugation in alkaline CsCl (pH 13,0). At this pH, DNA strands are separated, and any pieces of bacterial DNA would band at a heavy position if not covalently bonded to cellular DNA. The results of this experiment, represented in Fig. 3, show that no radioactivity appeared at the buoyant density of denatured *E.coli* DNA in alkaline CsCl, and suggest that bacterial and cellular DNAs are covalently linked.

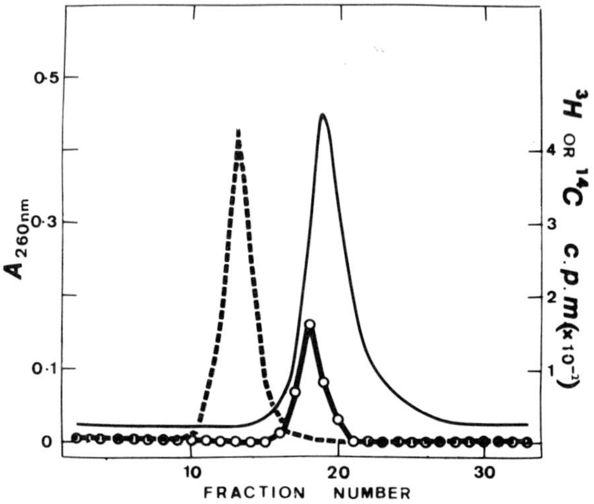

Fig. 3 - Centrifugation patterns in alkaline CsCl of NCTC cells lysed after 120 min of incubation with [^{14}C]-E.coli DNA. [^{3}H] - denatured E. coli DNA was added as a position marker. ——— absorbance; o——o [^{14}C] cts/min, ----- [^{3}H] cts/min.

(c) Role of cellular DNA synthesis

In order to correlate the mechanism of integration of exogenous DNA with the cell cycle and more precisely with the S. phase, two kinds of experiments were performed. In the first one, uptake and integration of bacterial DNA were followed in recipient cells where DNA synthesis was inhibited by hydroxyurea. In the other experiments, the separation of the newly synthesized DNA was obtained by incorporation of the heavy analog 5-BrUdr in the recipient cells, at the time of uptake of bacterial [^{14}C]-DNA.

Fate of E.coli DNA taken up under inhibition of cellular DNA synthesis

NCTC cells were grown in the presence of hydroxyurea (10^{-3}M) in order to assess the role of cellular DNA synthesis on the fate of incorporated $[^{14}C]$ - E. coli DNA. The residual DNA synthesis was 6% of the normal value without hydroxyurea, but the uptake of bacterial DNA was equivalent in normal or inhibited cells. However, the bacterial DNA was not integrated in the cellular DNA, as shown by CsCl density gradients. The distribution of $[^{14}C]$ among fractions of these gradients obtained from hydroxyurea inhibited cells is represented in Table III.

TABLE III

PARTITION OF INCORPORATED RADIOACTIVITY AMONG FRACTIONS OF E. COLI AND NCTC DNA BUOYANT DENSITIES, IN CsCl GRADIENTS PERFORMED ON CELLS INCUBATED WITH E. COLI DNA IN THE PRESENCE OF HYDROXYUREA

Incubation time (min)	% of radioactivity		
	E.coli DNA density	NCTC cell DNA density	Top of the gradient
0	0	0	0
30	75	0	0
45	81	2	0
60	71	3	18
90	20	4	70
120	0	6	88

Results are expressed in per cent of the maximum value observed after 120 min of incubation without inhibition of cellular DNA synthesis.

As in control cultures (Table I), E.coli DNA was taken up as a native molecule. Only 6% of the radioactivity appeared in the cellular DNA after 120 min of incubation, but the main part of foreign DNA was degraded in fragments which did not band in the gradient.

Growth in 5-BrUdr after uptake of E. coli DNA

Cells were incubated in Hanks solution containing $[^{14}C]$-E.coli DNA (1 μg/ml) and 5-BrUdr (10^{-5}M) for 120 min (Fig. 4 b), then centrifuged in neutral CsCl. Other cell cultures were incubated with $[^{14}C]$ -DNA and 5-BrUdr for 120 min, followed by incubation with 5-BrUdr for 2h (Fig. 4,c) or 4h (Fig. 4,d) before centrifugation. Fig. 4(a) shows the gradient pattern of recipient cells incubated with E.coli DNA for 2h without 5-BrUdr. The radioactivity due to incorporated foreign DNA was found in the heavier part of the cellular DNA. The buoyant density of the radioactive fractions increased with the incubation time of the recipient cell in the 5-BrUdr containing solution.

As in the experiments with hydroxyurea, this result shows that E.coli DNA is integrated during the synthesis of cellular DNA, and furthermore suggests that this synthesis was not interrupted by the integration of foreign DNA.

The centrifugation of the radioactive fractions corresponding to cells incubated with $[^{14}C]$ - DNA for 2h then with 5-BrUdr for 6h(Fig. 4,d) in alkaline CsCl, allowed the separation of the two strands of recipient DNA, the newly synthetized strand being made heavier by 5-BrUdr. The alkaline centrifugation shows that the integration takes place on both strands of the cellular DNA molecule, since radioactivity was found on the heavy and light

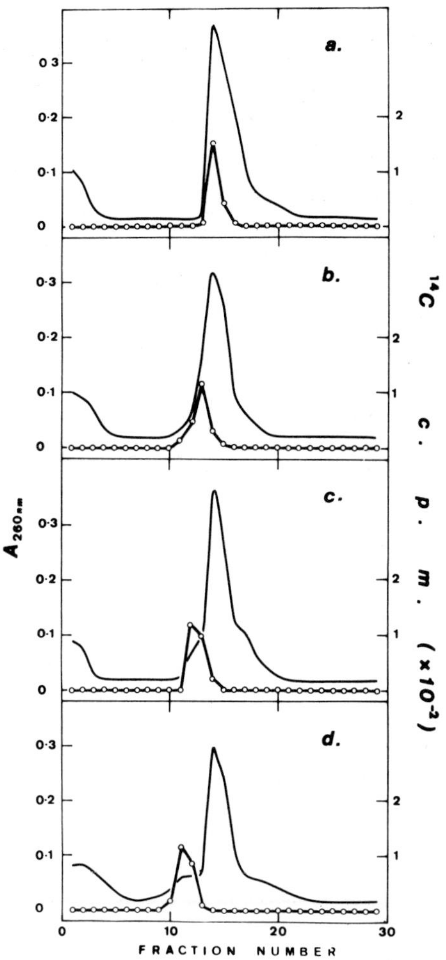

Fig. 4 - Neutral CsCl density gradient of a lysate of NCTC cells incubated for 120 min with [^{14}C] - E. coli DNA, then with 5-BrUdr for 0 (a), 2h (b), 4h (c) and 6h (d) ——— absorbance; 0——0 [^{14}C] cts/min.

strands of recipient DNA (LAVAL, MALAISE, to be published).

(d) Integration of bacterial DNA and repair synthesis

NCTC cells were grown in the presence of 5-BrUdr(10^{-5}M) for 8h, then in the presence of thymidine for 2h. Cells were X-irradiated,and incubated with $[^{3}H]$ - thymidine (5 μc/ml) or with $[^{14}C]$ - E. coli DNA (1 μg/ml) for 2h. After lysis, the DNA was analysed by preparative pycnography. The results are represented in Fig. 5. When NCTC cells were irradiated with doses of 10.000 rads, then incubated with $[^{3}H]$ - thymidine, a repair replication

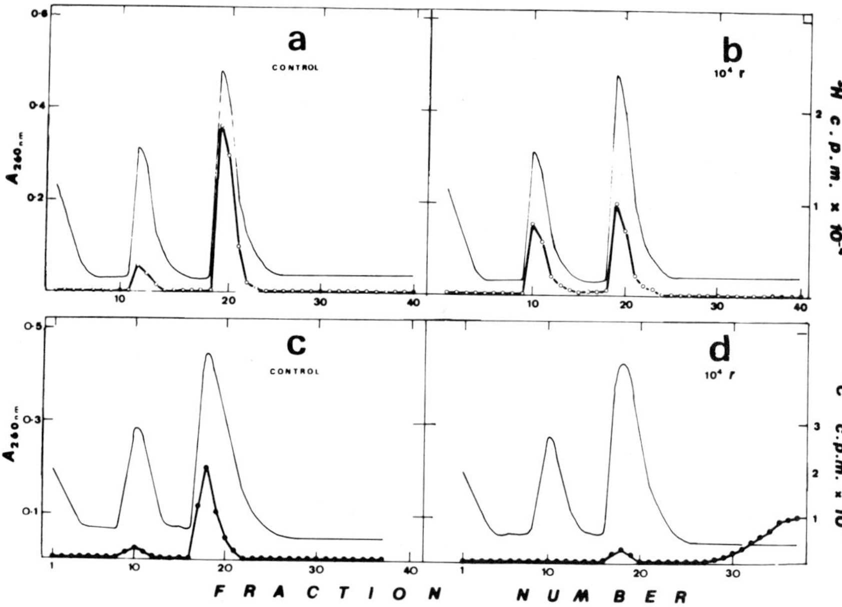

Fig. 5 - Equilibrium CsCl density gradients of DNA from NCTC cells grown in the presence of BrUdr. Cells were incubated with $[^{3}H]$ - thymidine for 120 min after irradiation with 0 (control, a) or 10.000 rads (b);or cells were incubated with $[^{14}C]$ -E.coli DNA for 120 min after irradiation with 0 (control, c) or 10.000 rads (d).

—— absorbance; ●——● $[^{14}C]$cts/min; o——o $[^{3}H]$cts/min.

could be detected.This was shown by the increase of the specific activity of the BU - containing DNA (Fig. 5 a and b), corresponding to cells which were out of the S-phase at the time of irradiation, and by the incorporation of the $[^{3}H]$ label on each of the two strands of this DNA (LAVAL & MALAISE, to be published). A depression of DNA synthesis was also observed, as shown for other mammalian cells in culture. (PAINTER & YOUNG, 1971).

When these irradiated cells were incubated with $[^{14}C]$-<u>E.coli</u> DNA,the uptake of exogenous DNA by the cells was the same than in control cultures (Fig. 5 c and d). However, its fate was different: no radioactivity was found in the "repaired" DNA, 5% of the incorporated radioactivity was linked to the cellular DNA corresponding to dividing cells, and 95% of the exogenous DNA taken up by the cells was hydrolyzed and was found at the top of the gradients.

These results correlate the previous experiments,since integration of <u>E.coli</u> DNA was reduced when cellular DNA synthesis was depressed by irradiation. They also show that the "repair synthesis"was not implicated in the integration of bacterial DNA in NCTC cells, after large doses of irradiation.

4. DISCUSSION

After the penetration of <u>E.coli</u> DNA into NCTC cells its fate could be divided in two steps. During the first one, the exogenous DNA penetrates into the cell nucleus and is in a double-stranded form, as judged by its buoyant density in CsCl gradient.This uptake was an energy-requiring process and was mainly influenced by the physical state of the foreign DNA. No foreign DNA could be detected in the mitochondrial fraction of the recipient cell. (LAVAL, MALAISE, LAVAL, 1970).

Then, a time dependent incorporation of the label corresponding to the foreign DNA occured into the cellular genome. All the radioactivity was linked to the host DNA 120 min after the beginning of uptake. Hybridization experiments and density gradient analysis showed that the foreign DNA was not degraded in the recipient cells, and that pieces of bacterial DNA were covalently linked to the cellular DNA,on each of the two strands of the recipient molecule.

The integration of donor DNA involved the cellular DNA synthesis,as shown by use of hydroxyurea and cell culture in the presence of 5-BrUdr. Experiments with X-irradiated cells showed that the repair synthesis (PAINTER & CLEAVER, 1969), occuring after doses of 10.000 rads, was not involved in the integration of foreign DNA.

These results suggested the possible sequence of events: (1) penetration of bacterial DNA into the cell nucleus by an energy-dependent process,the foreign DNA being on a double stranded form; (2) covalent linkage of these fragments to the DNA of recipient cells during the S-phase of the cell cycle. In this case, only donor DNA which has begun to interact specifically with the recipient DNA could be retained by the cell. This model would explain that uptake is irreversible (LAVAL, MALAISE, LAVAL 1971) and that no labeled nucleotides appeared in the recipient cells.

ACKNOWLEDGEMENTS

We would like to thank Prof. Maurice TUBIANA for his constant encouragement and advice.

5. REFERENCES

BHARGAVA, P. M. & SHANMUGAM, G. (1971). In Progress in Nucleic Acid Research and Molecular Biology, vol 11:103. Academic Press.

BURTON, K. (1956). Biochem. J. 62:315.

DENHARDT, D. J. (1966). Biochem. Biophys. Res. Commun. 23:641.

LAVAL, F., MALAISE, E. & LAVAL, J. (1970). Exp. Cell Res. 63:69.

LAVAL, F., MALAISE, E. & LAVAL, J. (1971). Informative Molecules in Biological Systems. L. Ledoux editor.124.

PAINTER, R. B. & CLEAVER, J. E. (1969). Radiat. Res. 37:451.

PAINTER, R. B. & YOUNG, B. R. (1971). Mutation Research 14:225.

SANFORD, K. K., LIKELY, G. D. & EARLE, W. R., (1954). J. Natl. Cancer Inst. 15:215.

SCHMIDT, G. & TANNHAUSER, S. J. (1945). J. Biol. Chem. 161:83.

VINOGRAD, J., MORRIS, J., DAVIDSON, N. & DOVE, W. F. (1963). Proc. Nat. Acad. Sci., Wash. 49:12.

AL-ANI, F. Y.
Department of Genetics, The University of Liverpool,
Brownlow Street, Liverpool L69 3BX, Great-Britain.

ANAGNOSTOPOULOS, Constantine
Centre de Génétique Moléculaire, Centre National de
la Recherche Scientifique,
91 Gif-Sur-Yvette, France.

ARCHER, Luís J.
Laboratório de Microbiologia, Instituto Gulbenkian
de Ciência,
Oeiras, Portugal.

AUDIT, Claudie
Centre de Génétique Moléculaire, Centre National de
la Recherche Scientifique,
91 Gif-Sur-Yvette, France.

BACH, Marie Louise
Institut de Botanique, Laboratoire de Génétique
Physiologique,
8, Rue Goethe, 67 Strasbourg, Germany.

BUTLER, Leslie O.
Department of Medical Microbiology, St. George's
Hospital Medical School, University of London,
Hyde Park Corner, London SW1, Great-Britain.

CARERE, Angelo*
Istituto Superiore di Sanitá,
Viale Regina Elena, 299, Roma, Italy.

CARMO, José Manuel
Laboratório de Microbiologia, Instituto Gulbankian
de Ciência,
Oeiras, Portugal.

CARPI, Natalia Guerola
Departamento de Genética, Facultad de Ciencias,
Universidad de Sevilla,
Sevilla, Spain.

CARVALHO, João Montezuma*
Instituto Botânico Dr. Júlio Henriques,
Universidade de Coimbra,
Coimbra, Portugal.

* Registered. Unable to participate.

CERDÁ, Enrique
Departamento de Genética, Facultad de Ciencias,
Universidad de Sevilla,
Sevilla, Spain.

CLAVERYS, Jean Pierre
Laboratoire de Génétique, Université Paul Sabatier,
118, Route de Narbonne, 31077 Toulouse Cedex, France.

COLE, Roger M.
Laboratory of Microbiology, National Institute of
Allergy and Infectious Diseases,
Bethesda, Maryland 20014, U.S.A.

COLLINS, John F.
Department of Molecular Biology, University of
Edinburgh,
King's Buildings, Mayfield Road,
Edinburgh EH9 3JT, Great-Britain.

CONCEIÇÃO, Izilda Costa*
Lombo de Carcavelos, lote 10,1º D.
Carcavelos, Portugal.

COSTA, João Vasconcelos
Laboratório de Biologia Celular, Instituto Gulbenkian
de Ciência,
Oeiras, Portugal.

COSTA, Maria Rosa
Department of Biology, Georgetown University,
Washington D.C. 20007, U.S.A.

DOBRZAŃSKI, Władysław A.
Department of Molecular Genetics, University Medical
School,
ul. Chalubinskiego 5, Warsaw, Poland.

DOLY, J.
Centre de Génétique Moléculaire, Centre National de la
Recherche Scientifique,
91 Gif-Sur-Yvette, France.

ELIZUR, Erela Ephrati*
Molecular Biology, Institute of Microbiology,
The Hebrew University, Hadassah Medical School,
Jerusalem, Israel.

* Registered. Unable to participate.

FOSSAS, Josefina Valoix*
Departamento de Microbiología, Facultad de Ciencias, Barcelona, Spain.

FRÖHOLM, Leif Oddvar
Methodology Department, National Institute of Public Health,
Geitmyrsveien 75, Oslo 1, Norway.

GARCIA, Rubens López
Instituto "Jaime Ferrán" de Microbiología, C.S.I.C.,
Joaquin Costa, 32, Madrid 6, Spain.

GLOVER, Stuart W.
Genetics Laboratory, Physics Building,
University of Newcastle,
Newcastle on Tyne NEI 7RU, Great-Britain.

GONÇALVES, Maria Helena
Instituto de Botânica "Dr. Gonçalo Sampaio",
Rua Campo Alegre, 1191, Porto, Portugal.

GOODGAL, Sol H.
Department of Microbiology, The School of Medicine,
University of Pennsylvania,
Philadelphia, Philadelphia 19104, U.S.A.

GROMKOVA, Rosa
Department of Microbiology, The School of Medicine,
University of Pennsylvania,
Philadelphia, Philadelphia 19104, U.S.A.

HARFORD, Nigel
Centre d'Etude de l'Energie Nucléaire,
Boeretang 200, B-2400 Mol, Belgium.

HERDMAN, Michael
Department of Biochemistry, University of Liverpool,
P.O.Box 147, Liverpool L69 3BX, Great-Britain.

HIROKAWA, Hideo
Max-Planck-Institut für Molekulare Genetik,
1 Berlin 33, Ihnestrasse 63-73, Germany.

KHOUDOKORMOFF, B.
Orange Plantage 38, Delft, Holland.

* Registered. Unable to participate.

KOHOUTOVÁ, Manci*
Institute of Microbiology, Czechoslovak Academy of Sciences,
Budejovická 270, Praha 4, KRC, Czechoslovakia.

KOPECKA, Helena
Institut de Biologie Moléculaire, Laboratory of Dr. Bernardi,
9, Quai Saint-Bernard, Paris 5e, France.

LANDMAN, Otto E.
Department of Biology, Georgetown University,
Washington D.C. 20007, U.S.A.

LAVAL, Françoise
Groupe de Recherche de Radiobiologie Clinique,
Institut Gustave Roussy,
94 Villejuif, France.

LEONARD, Carmen
Laboratory of Microbiology, National Institute of Allergy and Infectious Diseases,
Bldg. 5, Room 218, Bethesda, Maryland 20014, U.S.A.

LEONARD, Frederic*
Associate Program for Neurobiology,
National Science Foundation,
Washington D.C. 20550, U.S.A.

LEVIN, Barbara Chernov
Department of Biology, Georgetown University,
Washington D.C. 20007, U.S.A.

LEVIN, Ira W.
Department of Biology, Georgetown University,
Washington D.C. 20007, U.S.A.

LOSTIA, Onofrio*
Istituto Superiore di Sanitá,
Viale Regina Elena, 299, Roma, Italy.

MARSCH, Robert Cecil
Gesellschaft für Molekularbiologische Forschung mbH,
3301 Stöckheim über Braunschweig, Mascheroder Weg 1,
Germany.

* Registered. Unable to participate.

MENDES, Benilde Simões
Rua Dr. António Souto Lopes, 6 r/c D.,
Moscavide, Lisboa, Portugal.

MERGEAY, Max
Centre d'Etude de l'Energie Nucléaire,
Boeretang 200, B-2400 Mol, Belgium.

MOTA, Miguel
Estação Agronómica Nacional,
Oeiras, Portugal.

NESTER, Eugene W.
Department of Microbiology, University of Washington,
School of Medicine,
Seattle, Washington 98105, U.S.A.

PAKULA, Roman
Department of Microbiology, School of Hygiene,
University of Toronto,
Toronto 5, Ontario, Canada.

PEREZ, Evelio Perea*
Servicio de Microbiología, Clinica Puerta de Hierro,
San Martin de Porres, 4, Madrid 20, Spain.

PIGGOT, Patrick J.
University of Oxford, Microbiology Unit,
Department of Biochemistry,
South Parks Road, Oxford, Great-Britain.

POLSINELLI, Mario
Istituto di Genetica, Università di Pavia,
Via Sant'Epifanio, 14, 27100 Pavia, Italy.

PORTOLÉS, António
Instituto "Jaime Ferrán" de Microbiología,C.S.I.C.,
Joaquin Costa, 32, Madrid, Spain.

QUARTAU, Hermínia
Secção de Botânica, Faculdade de Ciências,
Universidade de Luanda,
Luanda, Angola.

RANHAND, Jon M.
NIAID-LM, Bldg. 5, Rm. 218, NIH,
Bethesda, Maryland 20014, U.S.A.

* Registered. Unable to participate.

RUTBERG, Lars*
Bakteriologiska Institutionem, Karolinska Institutet,
S-104 01 Stockholm 60, Sweden.

SAAVEDRA, Fernando Almeida
Rua António Cardoso, 556,
Porto, Portugal.

SCHUCH, Wolfgang*
Department of Zoology, University of Edinburgh,
West Mains Road, Edinburgh EH9 3JT, Great-Britain.

SICARD, Armand Michel
Centre de Recherche de Biochimie et de Génétique
Cellulaires, C.N.R.S.,
118 Route de Narbonne, 31077 Toulouse Cedex, France.

SIEWERT, Gerhard
Schering Ag., Postfach 650311,
1 Berlin 65, Germany.

TAPIA, Alfonso
Instituto "Jaime Ferrán" de Microbiología, C.S.I.C.,
Joaquin Costa, 32, Madrid 6, Spain.

TIRABY, Gérard
Department of Biology, Room 56-731,
Massachusetts Institute of Technology,
77 Massachusetts Avenue,
Cambridge, Massachusetts 02139, U.S.A.

TOMASZ, Alexander
The Rockefeller University,
66th. St. and York Avenue,
New York, N.Y. 10021, U.S.A.

TRAUTNER, Thomas A.
Max-Planck-Institut für Molekulare Genetik,
Ihnestrasse 63-73, 1 Berlin 33, Germany.

TROWSDALE, John*
Department of Genetics, The University of Birmingham,
P.O. Box 363, Birmingham B15 2TT, Great-Britain.

VENEMA, Gerard
Genetisch Instituut, Biologisch Centrum-Vleugel A,
Merklaan 30, Haren (Gn), Holland.

* Registered. Unable to participate.

VICENTE, Miguel
Departamento de Química Celular,
Centro de Investigaciones Biológicas,
Velazquez, 144, Madrid 6, Spain.

VIEIRA, Ana Maria
Rua Padre Antônio Vieira, 5, 4º,
Lisboa, Portugal.

VIERNY, Corinne
Centre de Génétique Moléculaire,
Centre National de la Recherche Scientifique,
91 Gif-Sur-Yvette, France.

YOUNG, Frank E.
Department of Microbiology, School of Medicine and
Dentistry and Strong Memorial Hospital,
260 Crittenden Boulevard,
Rochester, New York 14620, U.S.A.

ZILBER, Ilana*
Department of Microbiology, Tel-Aviv University,
155 Herzl Street, Tel Aviv, Israel.

* Registered. Unable to participate.